Monte Carlo Methods for Particle Transport

Second Edition

Monte Carlo Methods for Particle Transport

Second Edition

Alireza Haghighat

CRC Press
Taylor & Francis Group
Boca Raton London New York

CRC Press is an imprint of the
Taylor & Francis Group, an **Informa** business

Second edition published 2021
by CRC Press
6000 Broken Sound Parkway NW, Suite 300, Boca Raton, FL 33487-2742

and by CRC Press
2 Park Square, Milton Park, Abingdon, Oxon, OX14 4RN

First issued in paperback 2020

Library of Congress Cataloging-in-Publication Data
Names: Haghighat, Alireza, author.
Title: Monte Carlo methods for particle transport / Alireza Haghighat.
Description: Second edition. | Boca Raton : CRC Press, 2021. | Includes
    bibliographical references and index.
Identifiers: LCCN 2020017229 | ISBN 9780367188054 (hardback) | ISBN
    9780429198397 (ebook)
Subjects: LCSH: Monte Carlo method. | Particles (Nuclear
    physics)--Mathematical models. | Radiative transfer.
Classification: LCC QC20.7.M65 H34 2021 | DDC 518/.282--dc23
LC record available at https://lccn.loc.gov/2020017229

ISBN: 978-0-367-53809-5 (pbk)
ISBN: 978-0-367-18805-4 (hbk)
ISBN: 978-0-429-19839-7 (ebk)

DOI: 10.1201/9780429198397

Typeset in CMR
by Nova Techset Private Limited, Bengaluru & Chennai, India

To my wife, son, and mother.

Contents

CHAPTER 4 ▪ Fundamentals of Probability and Statistics   55

Acknowledgement

The second edition is a significant improvement over the first edition as it includes changes and additions in response to the feedback I received from students, practitioners, and reviewers over the past five years. Additionally, while using the book in my classes, I realized the need for revisions and additions. The second edition includes reorganized chapters, addition of new sections, especially in Chapters 4 and 10, deletion of redundant sections, and addition of the new Chapter 11 on the alternative eigenvalue Monte Carlo techniques to address the shortcoming of the standard techniques. I am hopeful that the new edition would benefit both students and practitioners, and promote further research and development on advanced Monte Carlo based techniques that not only are accurate, but also offer real-time simulation capabilities.

This new edition was not possible without valuable help I received from my students, colleagues and family. Particularly, I would like to express my highest gratitude to Valerio Mascolino, a PhD candidate in my group, who has provided highly valuable and careful review of the second edition. More specifically, he provided constructive suggestions, pointed out the need for clarification, identified errors and typos, helped with typing of the homework problems, fixed the references throughout the book, and prepared the book index. Also, I am thankful to Prof. William Walters, from Penn State, who provided constructive and highly valuable comments on Chapters 10 and 11. Finally, I should acknowledge that this book was greatly influenced by over three decades of research by my graduate students and investigators in the field, and by the inquisitive questions and comments of students.

It is important to acknowledge that the first edition was a 20-year project that dated back to my tenure at the Pennsylvania State University. As a newly minted assistant professor, I inherited an experimental course on Monte Carlo methods in particle transport in 1989 that had once been taught by the late Dr. Anthony Foderaro. The course was approved as a permanent part of Penn State's nuclear engineering curriculum in 1991. Three years later, in 1994, during my fifth year of teaching this course, I created the first bound version of my notes. The initial version of the notes relied heavily on Dr. Foderaro's unpublished notebook, and a number of other books and computer code manuals.

For the first edition of this book, three of my former graduate students, Prof. William Walters, Dr. Katherine Royston, and Dr. Nathan Roskoff, made valuable contributions on various aspects of the book. I remain truly grateful for their assistance and sincere interest in the preparation of the book. I am grateful to my colleagues and friends Prof. Bojan Petrovic, Prof. Farzad Rahnem, and Prof. Glenn Sjoden, Dr. Gianluca Longoni, and Dr. John Wagner who provided valuable reviews of the segments of the first edition.

My research group has been engaged in various research projects relating to the development of particle transport methodologies and codes for modeling and simulation of nuclear reactors, nuclear nonproliferation and safeguard detection systems, and radiation therapy and diagnostics systems. Specifically, my students and I have been involved in the development of automated variance reduction techniques for neutral and charged particle transport, and more recently hybrid methods for eigenvalue, and radiation detection and shielding calculations.

Lastly, I am highly grateful to my wife, Mastaneh, for her sacrifices and continuous care, support, and encouragement; my son, Aarash, who has always expressed interest and curiosity in my work, and has been a source of pride and inspiration; and my mother, Pari, who instilled in me a sense of achievement and integrity.

About the Author

Alireza Haghighat earned his PhD degree in nuclear engineering from the University of Washington, Seattle, in 1986.

Between 1986 and 2001, he was a professor of nuclear engineering at the Pennsylvania State University, University Park, Pennsylvania. From July 2001 to September 2009, he was Chair and Professor of the Department of Nuclear and Radiological Engineering at the University of Florida, Gainesville, Florida. From September 2009 to June 2011, Dr. Haghighat was a Florida Power & Light endowed term professor and served part-time as the Director of the University of Florida Training Reactor.

In January 2011, he joined the Mechanical Engineering Department at the Virginia Tech (VT) University, Greater Washington DC (GWDC) campus to help with the establishment of the VT Nuclear Engineering Program (VT-NEP) at both Blacksbirg and GWDC campuses. Currently, he is the Director of the NEP and also the Director of the Mechanical Engineering Graduate Program at GWDC.

Prof. Haghighat is a fellow of the American Nuclear Society (ANS). He leads the Center for Multiphysics for Advanced Reactor Simulation (MARS) and the Virginia Tech Theory Transport Group (VT^{3}G). Over the past 34 years, Prof. Haghighat has been involved in the development of new particle transport methodologies and large computer codes for modeling and simulation of nuclear systems including reactors, nuclear security and safeguards systems and medical devices. His efforts has resulted in the development of several advanced computer programs including PENTRAN, A^{3}MCNP, TITAN, INSPCT-s, AIMS, TITAN-IR, and RAPID. The latter four code systems are developed based on the novel Multi-stage Response-function Transport (MRT) methodology that results in simulation of nuclear systems in real time on one computer core. Additionally, for the RAPID code system, a virtual reality system (VRS) web application has been developed.

He has published over 250 papers, received several best paper awards, and presented many invited workshops, seminars, and papers nationally and internationally.

He is a recipient of the 2011 Radiation Protection Shielding Division's Professional Excellence Award, and received a recognition award from the Office of Global Threat Reduction for his leadership and contributions to design and analysis for the University of Florida Training Reactor HEU (highly enriched uranium) to LEU (low enriched uranium) fuel conversion in 2009.

Prof. Haghighat is an active member of ANS, and has served at various leadership positions, such as chair of the Reactor Physics Division, chair of the Mathematics and Computation Division, co-founder of the Computational Medical Physics Working Group, and chair of the NEDHO (Nuclear Engineering Department Heads Organization). Further, he is the founding Chair of the Virginia Nuclear Energy Consortium (VNEC) nonprofit organization.

Introduction

CONTENTS

The Monte Carlo method is a statistical technique which is capable of simulating a mathematical or physical experiment on a computer. In mathematics, it can provide the expectation value of functions and evaluate integrals; in science and engineering, it is capable of simulating complex problems which are comprised of various random processes with known or assumed probability density functions. To be able to simulate the random process, i.e., sample from a probability function for an event, it uses random numbers or pseudo-random numbers. Just like any statistical process, the Monte Carlo method requires repetition to achieve a small relative uncertainty, and therefore, may necessitate impractically large simulation times. To overcome this difficulty, parallel algorithms and variance reduction techniques are needed. This book attempts to address major topics affecting development, utilization, and performance of a Monte Carlo algorithm.

1.1 HISTORY OF MONTE CARLO SIMULATION

The birth of Monte Carlo simulation can be traced back to WWII. At that time, because of the Manhattan project, there was significant urgency in understanding nuclear fission and generating special nuclear materials. Great minds from all over the world were assembled in the United States to work on the Manhattan project. This coincided with another initiative: building of the first electronic computer. The first computer, ENIAC, had over 17,000 vacuum tubes in a system with

500,000 solder joints and was built at the University of Pennsylvania in Philadelphia under the leadership of physicist John Mauchly and engineer Presper Eckert [75]. The story is that, Mauchly was inspired to build an electronic computer to perform the work done in large rooms filled with mostly women calculating integrals for firing tables (ranges versus trajectories) for the U.S. Army. John von Neumann, who was a consultant for both the Army and Los Alamos National Lab (LANL) and was well aware of Edward Teller's new initiative in the area of thermonuclear energy, became interested in using ENIAC for testing models for thermonuclear reactions. He convinced the U.S. Army that it was beneficial for Los Alamos scientists (Nicholas Metropolis and Stan Frankel) to simulate thermonuclear reactions using ENIAC. The work started in 1945, before the end of the war, and its initial phase ended after the war in 1946. In addition to Metropolis, Frankel, and von Neumann, another scientist named Stanislaw (Stan) Ulam participated in a national project review meeting at LANL. It was Ulam's observation that the new electronic computer could be used for performing the tedious statistical sampling that was somewhat abandoned because of the inability to do large numbers of calculations. Von Neumann became interested in Ulam's suggestion and prepared an outline of a statistical approach for solving the neutron diffusion problem. Several people, including Metropolis, became interested in exploring the new statistical simulation approach; Metropolis suggested the name "Monte Carlo," which was inspired from the fact that Ulam's uncle used to borrow money from his family members because he "just had to go to Monte Carlo," a popular gambling destination in the Principality of Monaco. For the initial simulation, von Neumann suggested a spherical core of fissionable material surrounded by a shell of tamper material. The goal was to simulate neutron histories as they go through different interactions. To be able to sample the probability density functions associated with these interactions, he invented a pseudo-random number generator algorithm [104] referred to as middle-square digits,which was later replaced with more effective generators by H. Lehmer [64]. It was quickly realized that the Monte Carlo method was more flexible for simulating complex problems as compared to differential equations. However, since the method is a statistical process and requires the achievement of small variances, it was plagued by long times because of the need for a significant amount of computation! It is important to note that Enrico Fermi had already used the method for studying the moderation of neutrons using a mechanical calculator in the

Figure 1.1 Photo of FERMIAC (courtesy of Mark Pellegini, the Bradbury Museum, Los Alamos, NM)

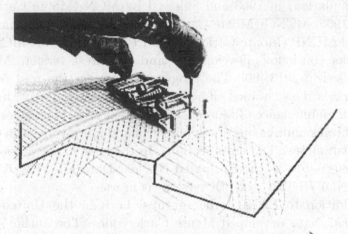

Figure 1.2 Application of FERMIAC (from N. Metropolis. 1987. Los Alamos Science 15: 125)

early 1930s while living in Rome. Naturally, Fermi was delighted with the invention of ENIAC; however, he came up with the idea of building an analog device called FERMIAC (shown in Figure 1.1) for the study of neutron transport. This device was built by Percy King and was limited to two energy groups, fast and thermal, and two dimensions. Figure 1.2 shows a demonstration of its application. Meanwhile, Metropolis and von Neumann's wife (Klari) designed a new control system for ENIAC to be able to process a set of instructions or a stored-program as opposed to the plugboard approach. With this new capability, Metropolis and von Neumann were able to solve several neutron transport problems. Soon other scientists in the thermonuclear group started studies with different geometries and for different particle energies. Later, mathematicians Herman Khan, C. J. Everett, and E. Cashwell became interested in the Monte Carlo method and published several articles on algorithms and use of the method for particle transport simulation [57, 56, 22, 33, 32].

1.2 STATUS OF MONTE CARLO CODES

The above brief history indicates that early development of Monte Carlo particle transport techniques were mainly conducted by the scientists at LANL. As a result, LANL has been the main source of general-purpose Monte Carlo codes, starting with MCS (Monte Carlo Simulation) in 1963 and followed by MCN (Monte Carlo Neutron) in 1965, MCNG (Monte Carlo coupled Neutron and Gamma) in 1973, and MCNP (Monte Carlo Neutron Photon) in 1977. MCNP has been under continuous development and the latest version, MCNP6, was released in 2013 [80]. The progress made over the past 50 years demonstrates the sustained effort at LANL on development, improvement, and maintenance of Monte Carlo particle transport codes. There has also been simultaneous development of and improvement in nuclear physics parameters, i.e., cross sections, in the form of the cross-section library referred to as the Evaluated Nuclear Data File (ENDF). Currently, ENDF/B-VIII, the 8th version, is in use.

In addition to LANL, other groups, both in the United States and abroad, have developed Monte Carlo codes. The author's group has developed the A^3MCNP (Automated Adjoint Accelerated MCNP) code system [43] for automatic variance reduction using the CADIS (Consistent Adjoint Driven Importance Sampling) methodology [46]. The Oak Ridge National Lab (ORNL) has developed a number of codes, including MORSE [31] and KENO [82], and more recently the ADVANTG code system [78] that uses CADIS and its alternate, "forward" CADIS (FW-CADIS) [109], methodology. Internationally, there are two general codes, MCBEND [120] and TRIPOLI [13], and a few specialized codes, including EGS [50], GEANT [1], and PENELOPE [89]. GEANT, which is an open-source code, has been developed in support of nuclear physics experiments. The other two codes, EGS and PENELOPE, have special focus on electron-photon transport for medical applications.

Finally, it is important to mention that in the past two decades there have been efforts on the development of codes based on the hybrid deterministic-Monte Carlo techniques. Chapter 11 is devoted to this topic.

1.3 MOTIVATION FOR WRITING THIS BOOK

Until the early 1990s, Monte Carlo methods were mainly used for benchmarking studies by scientists and engineers at national labs who had access to advanced computers. This situation, however, changed drastically with the advent of high performance computers (with fast clock cycles), parallel computers, and, more recently, PC clusters. To this effect, at the 8th International Conference on Radiation Shielding in 1994, over 70% of papers utilized deterministic methods for simulation of different problems or addressed new techniques and formulations. Since the late 1990s, this situation has been reversed, and Monte Carlo methods have become the first or, in some cases, only tool for performing particle transport simulations for a variety of applications. On the positive side, the method has enabled numerous people with different levels of knowledge and preparation to be able to perform particle transport simulations. However, on the negative side, it has created the potential of drawing erroneous results by novice users who do not appreciate the limitations of the method and/or the statistical concepts, and, therefore, cannot distinguish the difference between statistically reliable or unreliable results.

This book has been written to address these concerns and to provide relatively detailed discussions on the fundamental concepts and issues associated with the Monte Carlo method. Although this book is meant for engineers and scientists who are mainly interested in using the Monte Carlo method, the author has provided the necessary mathematical derivations to help understand issues associated with current techniques, and potentially explore new techniques. This book should be helpful to those who would like to simply learn about the method as well as those engaged in research and development.

The second edition of the book focuses on the correction and enhancement of mathematical derivations and associated physics and analysis. A number of chapters have been reorganized by adding or streamlining some of the discussions, removing redundant information, resulting in major modifications to Chapters 4 and 10, and the addition of the new Chapter 11. Chapter 11 addresses recently developed alternative eigenvalue methodologies that significantly improve the accuracy and efficiency of the method. The second edition, in addition to students, should be beneficial to the analysts engaged in the simulation of nuclear systems, and to those who are interested in the development of advanced techniques.

1.4 AUTHOR'S MESSAGE TO INSTRUCTORS

The author recommends this book as a textbook for graduate students in science and engineering. Over the past 31 years, the author has used the material presented in this book in his teaching of a course on the Monte Carlo methods to a diverse group of graduate students coming from various disciplines especially nuclear, computer science, electrical, mechanical and civil engineering, and physics. This valuable experience has helped significantly in shaping the organization and the content of both first and second editions of this book.

The book is effective for teaching to a diverse group of students, as it introduces the fundamentals of the method and issues related in the first five chapters. Chapter 6 gives an introduction to a simplified 1-D, one-speed neutron transport model and develops a simplified Monte Carlo particle transport algorithm. This topic is relatively easy to understand by a diverse group of students, and provides the means for learning issues related to the Monte Carlo method in general. Chapters 7-12 focus on various topics including variance reduction techniques, tallying, geometry, eigenvalue calculations and advanced techniques, and parallel and vector computing. Although Chapters 6 to 12 mainly focus on particle transport applications, their subject matters and supportive mathematical and statistical formulations can be pertinent to other applications.

The author believes that the students' learning would be enhanced significantly if they develop computer programs that can effectively apply the concepts and shed light on subtle issues of the Monte Carlo method.

In conclusion, the author hopes that the revisions and additions included in the second edition would make the book more beneficial to students, practitioners, and those engaged in the advancement of the Monte Carlo methods.

CHAPTER 2

Random Variables and Sampling

CONTENTS

2.1 INTRODUCTION

All basic (elementary) physical processes appear to be random; that is, one cannot predict, with certainty, what is the outcome of every individual process. Nevertheless, such random (stochastic) processes can generally be characterized by their average behavior and associated statistical uncertainties.

Outcomes of a physical process can be discrete or continuous. In other words, they can be selected (sampled) from a discrete or continuous event space. To be able to sample the outcome of a random process on a computer, it is necessary to identify the possible outcomes (random variables) and their types and probabilities, generate random numbers, and obtain a formulation between the random variables and random numbers. Commonly, the solution to this formulation is not straightforward therefore different methodologies have been developed for solving different types of equations. Significant efforts have been devoted to the development of methods [104, 33, 32, 58, 94, 96] that are very efficient, because numerous sampling is necessary for achieving a reliable average.

This chapter discusses different random variables and their probability functions, and derives a fundamental formulation relating random variables to random numbers. This formulation is referred to as the fundamental formulation of Monte Carlo (FFMC), as it provides the necessary formulation for performing a Monte Carlo simulation on a computer. Further, this chapter introduces different techniques for solving the FFMC formulation, and presents efficient solution techniques for a few commonly used distributions/functions.

2.2 RANDOM VARIABLES

Normally, outcomes are mapped onto numerical values for mathematical treatment. These numerical values are called random variables. So, just like an outcome, a random variable can be discrete or continuous. The outcome of tossing a die can be represented by a discrete random variable, while the time between particles emitted from a radioactive material can be represented by a continuous random variable.

For any random variable (x), two functions are defined: the probability density function (pdf) and the cumulative distribution function (cdf). In the following sections, these functions are described for discrete and continuous random variables.

2.2.1   Discrete random variable

The probability density function $(pdf, p(x_n))$ is the probability that the outcome of a random process is x_n. For example, for a well-balanced cubical die, the probability of any event x_n, is given by

$$p(x_n) = \frac{1}{6}, \ for \ n = 1, 6 \tag{2.1}$$

Note that the *pdf* is normalized such that the probability of getting any of the possible outcomes is exactly one (unity).

The cumulative distribution function $(cdf, P(x))$ is the probability that the outcome (random variable) of the random process has a value not exceeding x_n. For example, for the cubical die, we have

$$p(x_n) = \sum_{i=1}^{n} p(x_i) \tag{2.2}$$

Figure 2.1 shows the *pdf* and Figure 2.2 shows *cdf* for a well-balanced die.

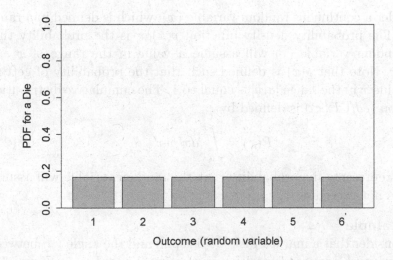

Figure 2.1   *pdf* associated with the "die" example (discrete random variable)

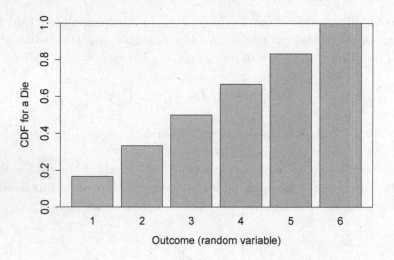

Figure 2.2   *cdf* associated with the "die" example (discrete random variable)

2.2.2   Continuous random variable

Consider a continuous random variable (x), which is defined in a range $[a, b]$. The probability density function $p(x)dx$ is the probability that the random variable (x) will assume a value in the range of x and $x + dx$. Note that $p(x)$ is defined such that the probability of getting any value x in the range $[a, b]$ is equal to 1. The cumulative distribution function $(cdf, P(x))$ is defined by

$$P(x) = \int_a^x dx' p(x') \tag{2.3}$$

which represents the probability that the random variable will assume a value not exceeding x.

Example
Consider that a marked disk [35] is spun and the angle (ϕ) between the mark and a reference position is measured, as shown in Figure 2.3. If this process is repeated, a different value ϕ is obtained every time. This means that the process is a random process, and ϕ is a continuous random variable varying in a range of $[0, 2\pi]$.

What is the pdf for this process? For a well balanced disk, the probability of getting any ϕ in the range $d\phi$ should be independent of

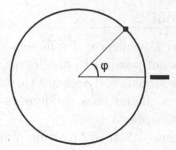

Figure 2.3   Schematic of a disk experiment for generation of a continuous random number

ϕ, i.e., constant. We may define a probability density function $(p(\phi))$ given by

$$p(\phi) = k \tag{2.4}$$

The constant (k) is obtained by setting a normalization requirement on the *pdf* as follows

$$\int_0^{2\pi} d\phi p(\phi) = 1$$

$$\int_0^{2\pi} d\phi k = 1 \tag{2.5}$$

$$2\pi k = 1$$

$$p(\phi) = k = \frac{1}{2\pi}$$

The *cdf* is given by

$$P(\phi) = \int_0^{\phi} d\phi' p(\phi') = \frac{\phi}{2\pi} \tag{2.6}$$

2.2.3   Notes on *pdf* and *cdf* characteristics

The *pdf* and *cdf* of a random variable have a few important characteristics as summarized below:

1. *pdf* is always positive.

2. *cdf* is always positive and nondecreasing function of its random variable.

3. *pdf* is normalized such that its corresponding *cdf* varies in a range of $[0, 1]$.

2.3 RANDOM NUMBERS

Random Numbers are: a sequence of numbers that have the special characteristic that it is not possible to predict η_{n+1} based on the previous η_n numbers in the sequence. To assure this unpredictability (randomness), the numbers should pass randomness tests. This topic is discussed in detail in Chapter 3.

To generate a sequence of (truly) random numbers, we need a generating approach (function) that yields numbers that are uniformly distributed in a range (commonly, a range of [0, 1]).

Example

The disk experiment in the previous section can be used to generate a sequence of random numbers: spin the disk to get ϕ, evaluate *cdf* ($= \frac{\phi}{2\pi}$), and set $\eta = \frac{\phi}{2\pi}$. This experiment is a good random number generator, because η is a random variable as ϕ is a random variable, and its value is in the range of [0, 1].

What is the *pdf* for generating random numbers $\eta's$? To preserve the desired characteristics mentioned above, the *pdf* for a random number generator is given by

$$q(\eta) = 1, \ \ for \ 0 \leq \eta \leq 1 \tag{2.7}$$

Therefore, the corresponding *cdf* reads as

$$Q(\eta) = \int_0^\eta d\eta' q(\eta'), \ \ for \ 0 \leq \eta \leq 1 \tag{2.8}$$

Now, let's derive the *pdf* and *cdf* for the random number ($= \frac{\phi}{2\pi}$) generated via the disk experiment. Consider that

$$\eta = \frac{\phi}{2\pi} \tag{2.9}$$

Because there is a one-to-one relation between η and ϕ, knowing $p(\phi)$, we can obtain a *pdf* for η using the following equality

$$|q(\eta)d\eta| = |p(\phi)d\phi| \tag{2.10}$$

Since $q(\eta)$ and $p(\phi)$ are positive functions, we solve for $q(\eta)$ using the following formulation

$$q(\eta) = p(\phi)|\frac{d\phi}{d\eta}| = p(\phi)(2\pi) \tag{2.11}$$

Substituting for $p(\phi) = \frac{1}{2\pi}$, $q(\eta)$ reduces to

$$q(\eta) = (\frac{1}{2\pi})2\pi = 1, \ for \ 0 \leq \eta \leq 1 \qquad (2.12)$$

and the corresponding *cdf* is given by

$$Q(\eta) = \int_0^\eta d\eta' q(\eta'), \ for \ 0 \leq \eta \leq 1 \qquad (2.13)$$

Therefore, we conclude that the random numbers generated via the disk using the *cdf* formulation for the random variable (ϕ) satisfies the desired conditions set forth for generating random numbers. Further discussion on random number generation is given in Chapter 3.

2.4 DERIVATION OF THE FUNDAMENTAL FORMULATION OF MONTE CARLO (FFMC)

So far, we have discussed random variables and random numbers. In a Monte Carlo simulation, the goal is to simulate a physical process in which we are knowledgeable about the basic physics, i.e., we know the *pdf*s of the basic processes. Assuming that we can generate the random numbers, we want to obtain the random variable (x), i.e., sample the outcome of the random process x with $p(x)$. We consider that the random variable (x) is related to the random number (η), hence, we may write

$$p(x)dx = q(\eta)d\eta, \qquad for \ \ a \leq x \leq b \ and \ 0 \leq \eta \leq 1 \qquad (2.14)$$

Then, we may integrate both sides of above equation in the range $[a, x]$ and $[0, \eta]$ to obtain

$$\int_a^x dx' p(x') = \int_0^\eta d\eta' \cdot 1$$
$$P(x) = \eta \qquad (2.15)$$

Equation 2.15 gives a relation for obtaining a continuous random variable x using a random number η. This relation is illustrated in Figure 2.4. The remaining question is: How do we deal with discrete random variables? Because the random number η is a continuous variable, while a discrete *pdf*, $P(n)$, only assumes certain values, we have to set the following relation

$$Min[P(n)|P(n) \geq \eta] \qquad (2.16)$$

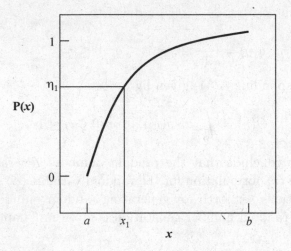

Figure 2.4 Sampling a continuous random variable x

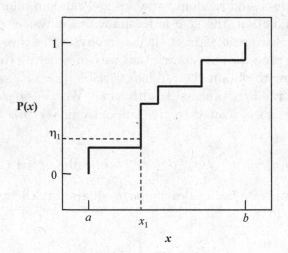

Figure 2.5 Sampling a discrete random variable x_i

where, $P(n) = \sum_{i=1}^{n} p_i$. This means that n is selected when the minimum of $P(n)$ is greater than or equal to η. This relation is illustrated in Figure 2.5.

2.5 SAMPLING ONE-DIMENSIONAL DENSITY FUNCTIONS

In this section, we discuss different approaches [58, 94, 35] for solving the FFMC for processes with a single random variable, i.e., with one-dimensional density functions.

2.5.1 Analytical inversion

The FFMC is inverted to obtain a formulation for a random variable x in terms of a number $\eta \in [0,1]$. Mathematically, this means that we obtain an inverse formulation, $x = P^{-1}(\eta)$. As an example, if the *pdf* for the random variable x is given by

$$p(x) = \frac{1}{2}, \quad for \; -1 \leq x \leq 1 \tag{2.17}$$

then, the corresponding FFMC formulation is given by

$$\int_{-1}^{x} dx' \frac{1}{2} = \eta \tag{2.18}$$

and, x or $P^{-1}(\eta)$ is given by

$$x = 2\eta - 1. \tag{2.19}$$

As the *pdf* becomes complex, obtaining $P^{-1}(\eta)$ analytically becomes more complicated or even impossible; therefore, other techniques are needed.

2.5.2 Numerical inversion

If the analytical inversion is impractical or impossible, then we may follow a numerical procedure. The *pdf* is partitioned into N equiprobable areas within $[a, b]$. This means that each area is equal to

$$\int_{x_{i-1}}^{x_i} dx' p(x') = \frac{1}{N} \tag{2.20}$$

Figure 2.6 demonstrates this process. Then, the average *pdf* within each interval is given by

$$p_i = \frac{1}{N(x_i - x_{i-1})} \quad for \; i = 1, N \tag{2.21}$$

The first step is to determine $x_i s$ by using the following procedure presented in Table 2.1.

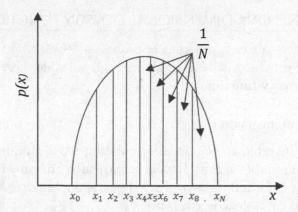

Figure 2.6 Demonstration of partitioning of a density function into N equal areas

Table 2.1 A procedure for partitioning a pdf into N equal areas

Set N (*Number of equiprobable areas*)
do $i = 1, N$
    $p_i = p(x_{i-1})$
    $x_i = x_{i-1} + \frac{1}{Np_i}$ (*Using Equation 2.21*)
    $area = area + p_i(x_i - x_{i-1})$
enddo
relative-difference (%)=$100 * abs(area - 1)$
if(relative-difference $\leq \varepsilon$) then
    N is adequate
else
    increase N, repeat the process.
endif

Note that the above procedure is done one time for a given pdf. After this, we can sample the pdf using a two-step sampling process as follows:

1. Generate two random numbers (η_1 and η_2).

2. Sample an interval (area) using

$$i = INT(N \cdot \eta_1) + 1. \tag{2.22}$$

3. Sample x within the ith interval using

$$x = x_{i-1} + \eta_2(x_i - x_{i-1}).\qquad(2.23)$$

2.5.3 Probability mixing method

If the *pdf*, $p(x)$, can be partitioned into n nonnegative functions, i.e.,

$$p(x) = \sum_{i=1}^{n} f_i(x), \quad for \ f_i(x) \geq 0 \ and \ a \leq x \leq b,\qquad(2.24)$$

then, we may define *pdf* s corresponding to each $f_i(x)$ as

$$p_i(x) = \alpha_i f_i(x),\qquad(2.25)$$

where α_i is a constant required to normalize f_i. Hence, the $p(x)$ formulation reduces to

$$p(x) = \sum_{i=1}^{n} \frac{1}{\alpha_i} p_i(x).\qquad(2.26)$$

Considering that the sum of coefficients $\left(\sum_{i=1}^{n} \frac{1}{\alpha_i}\right)$ is equal to 1, we devise a two-step procedure for sampling the random variable x as follows:

1. Generate a RN (η_1), and then select the $i^{th} pdf$, $p_i(x)$, using the following inequality

$$\sum_{i'=1}^{i-1} \frac{1}{\alpha_{i'}} < \eta_1 \leq \sum_{i'=1}^{i} \frac{1}{\alpha_{i'}},\qquad(2.27)$$

   where $\frac{1}{\alpha_i} = \int_a^b dx f_i(x)$.

2. Generate a RN (η_2), and then sample x from the selected $i^{th} pdf$, $p_i(x)$, by

$$\eta_2 = P_i(x) = \alpha_i \int_a^x dx' f_i(x').\qquad(2.28)$$

Note that this method is only useful if Equation 2.28 is easily solvable, i.e., each individual $f_i(x)$ should be analytically invertible.

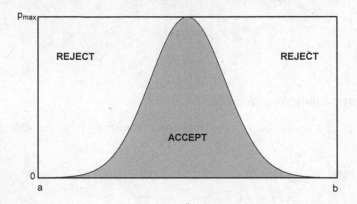

Figure 2.7 Demonstration of the rejection technique

2.5.4 Rejection technique

If exact computation of $P^{-1}(\eta)$ is not straightforward, we may consider the rejection technique as follows:

1. Enclose $p(x)$ in a frame bounded by p_{max}, a, and b, as shown in Figure 2.7.

2. Generate two random numbers η_1 and η_2.

3. Sample the random variable x using

$$x = a + \eta_1(b - a). \tag{2.29}$$

4. Accept x if

$$\eta_2 p_{max} \leq p(x). \tag{2.30}$$

Note that in this technique all the pairs $(x, y = \eta_2 p_{max})$ are accepted if they are bounded by $p(x)$, otherwise, they are rejected. So, effectively, we are sampling from the area under the *pdf*, i.e., *cdf*. Because the technique samples from the area, it is straightforward to define an efficiency formulation given by

$$efficiency = \frac{\int_a^b dx p(x)}{p_{max}(b - a)} = \frac{1}{p_{max}(b - a)} \tag{2.31}$$

This technique is simple to implement and effective, but if the efficiency is low, then it can be very slow.

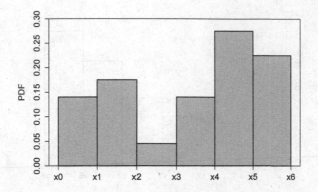

Figure 2.8   A histogram representation of a *pdf*

2.5.5   Numerical evaluation

If a continuous *pdf* is represented as a histogram, as shown in Figure 2.8, then its *cdf* is obtained using

$$P_i = \sum_{i'=1}^{i} p_{i'}(x_{i'} - x_{i'-1}), \quad for \ \ i = 1, n. \tag{2.32}$$

Figure 2.9 shows a representation of P_i.

To obtain a FFMC for a continuous random variable, we develop an interpolation formulation given by

$$P(x) = \frac{x - x_{i-1}}{x_i - x_{i-1}} P_i + \frac{x_i - x}{x_i - x_{i-1}} P_{i-1}. \tag{2.33}$$

Then the FFMC for x reads as

$$\eta = \frac{x - x_{i-1}}{x_i - x_{i-1}} P_i + \frac{x_i - x}{x_i - x_{i-1}} P_{i-1} \tag{2.34}$$

Therefore, x is sampled as

$$x = \frac{(x_i - x_{i-1})\eta - x_i P_{i-1} + x_{i-1} P_i}{P_i - P_{i-1}} \tag{2.35}$$

To implement this method, the following procedure is used:

1. A random number η is generated.

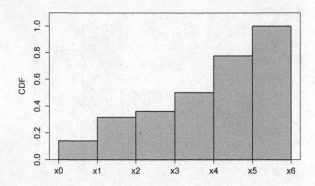

Figure 2.9 A *cdf* corresponding to a *pdf* represented in a histogram format

2. A search for i is conducted such that $P_{i-1} < \eta \le P_i$.

3. Random variable x is evaluated by Equation 2.35.

The second step in this procedure can use either linear or binary algorithms. In a linear algorithm, the search is performed by marching monotonically from the minimum value to the maximum value of the *cdf* until the necessary condition of the random variable is satisfied. In a binary search, the following algorithm is used:

1. Generate a random variable η.

2. The value of the *cdf* of the mid-number from the list of *cdfs* is identified.

3. A half sequence is selected:

   (a) Lower-half sequence for $\eta > cdf(mid - number)$

   (b) Upper-half sequence for $\eta < cdf(mid - number)$

4. Iterate on steps 2 and 3, until the appropriate inequality for the random variable is satisfied.

It is important to note that computation times for the linear and binary searches are $O(N)$ and $O(log_2(N))$, respectively. Hence, for large values of N, the binary search can be significantly more efficient.

2.5.6   Table lookup

In this approach, a table of *cdf* versus random variable is created and stored in the computer memory. Then sampling is performed by generating a random number, which is compared with the *cdf* to determine the random variable. Depending on the type of random variable, i.e., continuous versus discrete, somewhat different procedures are used as follows. In the case of a continuous random variable, the table entries resemble a histogram; therefore, the procedure discussed in Section 2.5.5 should be applied.

In the case of a discrete random variable, the following procedure is used:

1. A random variable η is generated.

2. A search for i is conducted such that the $P_{i-1} < \eta \leq P_i$ inequality is satisfied.

Note that a linear or binary search algorithm, as discussed in the previous section, should be employed in step 2.

2.6   SAMPLING MULTIDIMENSIONAL DENSITY FUNCTIONS

In Section 2.5, we introduced different techniques for solving the FFMC formulation corresponding to a one-dimensional density function, i.e., a process with a single random variable. In this section, we discuss how to develop FFMC formulations for sampling random variables for a random process with more than one random variable [94]. Considering a general density function expressed by

$$f(x_1, x_2, x_3, ...x_n), \tag{2.36}$$

where $a_1 \leq x_1 \leq b_1, \ a_2 \leq x_2 \leq b_2, \ \cdots, a_n \leq x_n \leq b_n$.

Here, for this multidimensional density function, we derive one-dimensional probability density functions for sampling each random variable as follows:

Starting with the first random variable (x_1), the *pdf* is given by

$$p_1(x_1) = \frac{\int_{a_2}^{b_2} dy_2 \int_{a_3}^{b_3} dy_3 \int_{a_4}^{b_4} dy_4 \cdots \int_{a_n}^{b_n} dy_n f(x_1, y_2, y_3, \cdots, y_n)}{\int_{a_1}^{b_1} dy_1 \int_{a_2}^{b_2} dy_2 \int_{a_3}^{b_3} dy_3 \int_{a_4}^{b_4} dy_4 \cdots \int_{a_n}^{b_n} dy_n f(y_1, y_2, y_3, \cdots, y_n)}. \tag{2.37}$$

For the second random variable (x_2), the conditional *pdf*, given x_1, is given by

$$p_2(x_2|x_1) = \frac{\int_{a_3}^{b_3} dy_3 \int_{a_4}^{b_4} dy_4 \cdots \int_{a_n}^{b_n} dy_n f(x_1, x_2, y_3, \cdots, y_n)}{\int_{a_2}^{b_2} dy_2 \int_{a_3}^{b_3} dy_3 \int_{a_4}^{b_4} dy_4 \cdots \int_{a_n}^{b_n} dy_n f(x_1, y_2, y_3, \cdots, y_n)}. \tag{2.38}$$

For the third random variable (x_3), the conditional *pdf*, given x_1 and x_2, is given by

$$p_3(x_3|x_1, x_2) = \frac{\int_{a_4}^{b_4} dy_4 \cdots \int_{a_n}^{b_n} dy_n f(x_1, x_2, x_3, \cdots, y_n)}{\int_{a_3}^{b_3} dy_3 \int_{a_4}^{b_4} dy_4 \cdots \int_{a_n}^{b_n} dy_n f(x_1, x_2, y_3, \cdots, y_n)}. \tag{2.39}$$

Similarly, for the nth random variable (x_n), the conditional *pdf*, given $x_1, x_2, x_3, \ldots, x_{n-1}$, is given by

$$p_n(x_n|x_1, x_2, x_3, \cdots, x_{n-1}) = \frac{f(x_1, x_2, x_3, \cdots, x_n)}{\int_{a_n}^{b_n} dy_n f(x_1, x_2, y_3, \cdots, y_n)}. \tag{2.40}$$

Hence, the corresponding FFMCs for a multidimensional probability density function are given by

$$P_1(x_1) = \int_{a_1}^{x_1} dy_1 p_1(y_1) = \eta_1 \tag{2.41}$$

$$P_2(x_2) = \int_{a_2}^{x_2} dy_2 p_2(y_2|x_1) = \eta_2 \tag{2.42}$$

$$P_3(x_3) = \int_{a_3}^{x_3} dy_3 p_2(y_3|x_1, x_2) = \eta_3 \tag{2.43}$$

$$P_n(x_n) = \int_{a_n}^{x_n} dy_n p_n(y_n|x_1, x_2, \cdots, x_{n-1}) = \eta_n \tag{2.44}$$

Note that these formulations can be solved via the techniques discussed in Section 2.5 for single variable *pdf's*.

A simple example for a multidimensional *pdf* is the density function for selection of polar and azimuthal angles given by

$$p(\mu, \phi)d\mu d\phi = \frac{1}{4\pi} d\mu d\phi, \quad -1 \le \mu \le 1 \tag{2.45}$$

The corresponding *pdfs* for the two random variables are derived as follows

$$p_1(x_1) = \frac{\int_{a_2}^{b_2} dy_2 p(x_1, y_2)}{\int_{a_1}^{b_1} dy_1 \int_{a_2}^{b_2} dy_2 p(y_1, y_2)} \tag{2.46}$$

Considering, $x_1 = \mu$, and $x_2 = \phi$, then above equation reduces to:

$$p_1(\mu) = \frac{\int_0^{2\pi} d\phi p(\mu, \phi)}{\int_1^1 d\mu \int_0^{2\pi} d\phi p(\mu, \phi)};$$

$$p_1(\mu) = \frac{\int_0^{2\pi} d\phi \frac{1}{4\pi}}{1} = \frac{1}{2}; \tag{2.47}$$

and,

$$p_2(\phi|\mu) = \frac{p(\mu, \phi)}{\int_0^{2\pi} d\phi p(\mu, \phi)};$$

$$p_2(\phi|\mu) = \frac{\frac{1}{4\pi}}{\int_0^{2\pi} d\phi \frac{1}{4\pi}} = \frac{1}{2\pi}. \tag{2.48}$$

Therefore, the corresponding FFMCs are given by

$$P_1(\mu) = \int_{-1}^{\mu} dy_1 p_1(y_1) = \eta_1;$$

$$\int_{-1}^{\mu} \frac{1}{2} = \eta_1; \tag{2.49}$$

$$\mu = 2\eta_1 - 1,$$

and

$$P_2(\phi) = \int_0^{\phi} dy_2 p_2(y_2) = \eta_2;$$

$$\int_0^{\phi} dy_2 \frac{1}{2\pi} = \eta_2; \tag{2.50}$$

$$\phi = 2\pi\eta_2.$$

2.7 EXAMPLE PROCEDURES FOR SAMPLING A FEW COMMONLY USED DISTRIBUTIONS

As discussed earlier, the computation time of a Monte Carlo simulation is highly dependent on the procedure used for sampling from the different *pdf's* encountered. Hence, different researchers have worked on developing highly efficient algorithms. In this section, we will introduce algorithms proposed for a few functions encountered in particle transport problems.

2.7.1 Normal distribution

The normal distribution is commonly encountered in modeling most physical phenomena, and, therefore, various researchers [14, 58] have developed highly efficient methodologies for its sampling. One of the most effective approaches for sampling is referred to as the Box-Muller procedure [14].

We define the Box-Muller technique to sample a normal distribution given by

$$\phi(t) = \frac{1}{\sqrt{2\pi}} e^{-\frac{t^2}{2}}. \tag{2.51}$$

Consider two independent random variables x and y that follow normal distributions. The combined probability of these two variables is expressed by

$$\phi(x,y)dxdy = \frac{1}{2\pi} e^{-\frac{x^2+y^2}{2}} dxdy \tag{2.52}$$

If we consider (x, y) as components of the (x, y) frame of reference, we can express them in terms of polar coordinates (r, θ) as follows

$$\begin{aligned} x &= r cos\theta, \\ y &= r sin\theta, \end{aligned} \tag{2.53}$$

and differential area

$$dxdy = rdrd\theta. \tag{2.54}$$

Hence, the distribution function (Equation 2.54) can be expressed in polar coordinates as

$$f(r,\theta)drd\theta = \frac{1}{2\pi} e^{-\frac{r^2}{2}} rdrd\theta. \tag{2.55}$$

Note that the right-hand side of the above equation can be written as two independent density functions

$$f(r,\theta)drd\theta = \left[e^{-\frac{r^2}{2}} rdr \right] \left[\frac{d\theta}{2\pi} \right]. \tag{2.56}$$

Then, r and θ random variables can be sampled independently, and used to determine x and y random variables as outlined in Table 2.2. Note that the procedure in Table 2.2 indicates that, by generating two random numbers, one obtains two random variables (x and y).

Table 2.2 Procedure for sampling from a normal distribution

$$p(\theta) = \frac{1}{2\pi} \qquad \int_0^\theta d\theta \frac{1}{2\pi} = \eta_1 \qquad\qquad \text{Sample } \theta = 2\pi\eta_1$$

$$p(r) = re^{-\frac{r^2}{2}} \qquad \int_0^r drre^{-\frac{r^2}{2}} = \eta_2 \qquad\qquad \text{Sample } r = \sqrt{-2\ln\eta_2}$$

$$\text{Sample } x, y \qquad x = \sqrt{-2\ln\eta_2}\cos(2\pi\eta_1) \qquad y = \sqrt{-2\ln\eta_2}\sin(2\pi\eta_1)$$

2.7.2 Watt spectrum

This distribution is commonly used for sampling energy of fission neutrons, i.e., fission neutron spectrum. The distribution is expressed by

$$W(a, b, E') = Ae^{-aE'}sinh\sqrt{bE'}, \ \ 0 < E' < \infty, \qquad (2.57)$$

where

$$A = \frac{\left(\sqrt{\frac{\pi b}{4a}}\right)e^{\frac{b}{4a}}}{a} \qquad\qquad (2.58)$$

Here, a and b vary with the fissile isotope, and are weakly dependent on the energy of the incident neutron. An efficient procedure for sampling E' (fission neutron energy) is derived by [32] as outlined in Table 2.3.

2.7.3 Cosine and sine functions sampling

In particle transport simulations, sampling from the cosine and sine functions is often needed. Because the sine and cosine functions are

Table 2.3 Procedure for sampling from the Watt spectrum

Set $L = a^{-1}(k + \sqrt{k^2 - 1})$, where $k = 1 + \frac{b}{8a}$

sample x and y $\qquad\qquad x = -\ln\eta_1$ and $y = -\ln\eta_2$

If $(y - M(x + 1))^2 \leqslant bLx$,, where $M = aL - 1$

Set $E' = Lx$

computationally expensive, a more efficient approach was developed by [104] as follows.

Consider one quarter of a circle of radius 1 cm, where both x and y have positive values. Then, follow the procedure outlined in Table 2.4 to obtain the cosine and sine of an angle.

Table 2.4 Procedure for sampling from a sine or cosine function

Generate, η_1, η_2 x and y $\varepsilon [0,1]$		$x = \eta_1, y = \eta_2$
$cos\theta = \frac{x}{\sqrt{x^2+y^2}}$,	hence, $cos\theta = \frac{\eta_1}{\sqrt{\eta_1^2+\eta_2^2}}$	for $\theta \in \left[0, \frac{\pi}{2}\right]$
$sin\theta = \frac{y}{\sqrt{x^2+y^2}}$,	hence, $sin\theta = \frac{\eta_2}{\sqrt{\eta_1^2+\eta_2^2}}$	for $\theta \in \left[0, \frac{\pi}{2}\right]$

To sample from sine and cosine over the entire range, i.e., $\theta \in [0, 2\pi]$, von Neumann demonstrated the use of 2θ instead. For the cosine, since $\cos(\theta) = \cos(-\theta)$, the range of $[0, \pi]$ is sufficient. Using the formulations derived for θ, the cosine formulation for 2θ is given by

$$\cos(2\theta) = \cos^2\theta - \sin^2\theta = \frac{\eta_1^2}{\eta_1^2 + \eta_2^2} - \frac{\eta_2^2}{\eta_1^2 + \eta_2^2} = \frac{\eta_1^2 - \eta_2^2}{\eta_1^2 + \eta_2^2} \quad (2.59)$$

For the sine, since $\sin(\theta) = -\sin(-\theta)$, the range of $[-\pi, \pi]$ has to be considered. Using the formulations derived for θ, the sine formulation for $\pm 2\theta$ is given by

$$\sin(\pm 2\theta) = \pm 2\sin(\theta)\cos(\theta) =$$
$$\pm 2\frac{\eta_1^2}{\sqrt{\eta_1^2 + \eta_2^2}} \cdot \frac{\eta_2^2}{\sqrt{\eta_1^2 + \eta_2^2}} = \pm 2\frac{\eta_1^2\eta_2^2}{\eta_1^2 + \eta_2^2} \quad (2.60)$$

To sample positive and negative values, the following procedure is considered:

1. A random number, η_3, is generated;

2. If $\eta_3 \leq 0.5$, a positive sign is selected, otherwise a negative sign is selected.

2.8 REMARKS

Every random variable has two associated functions: a *pdf* and a corresponding *cdf*. Knowing these functions, one can predict the outcome of a random process through sampling from these distributions, as applicable to the task. To simulate a random process on a computer, we need to sample the associated random variables. To do so, normally we generate a set of pseudo-random numbers, which are used to obtain the random variable. This is accomplished by forming the FFMC, which provides a one-to-one relation between the random variable and a random number. Finally, it is demonstrated that, for each FFMC formulation, one may have to examine different techniques for achieving an unbiased solution in a very short time.

PROBLEMS

1. Consider a pair of well-balanced, six-sided dice:

   a. Diagram a flowchart for an algorithm to randomly select the sum of the top faces (n_1, n_2) of the pair based on generated random number η's.

   b. Write a program using the previous algorithm to estimate the *pdf* for the sum of two dice rolls $(s = n_1 + n_2)$. Run the program for 1,000 and 50,000 pairs of dice rolls. Compare the results to the true *pdf*.

2. Consider that a standard deck of cards has 52 cards, with 13 cards of each of 4 suits.

   a. Diagram a flowchart for an algorithm to randomly select a five-card poker hand based on generated random numbers.

   b. Write a program using the previous algorithm to estimate the probability of getting a flush. A flush hand is when all five cards are of the same suit (any suit). Run the program for 1,000 and 50,000 hands. Compare the results with the true probability.

3. Random variables x and y have a one-to-one relation as follows:

$$y = x^2 + 1, \qquad \text{for} \quad 0 \le x \le 1$$

determine the *pdf* for the random variable x, given:

$$f(y) = y + 1$$

4. Consider a continuous random variable x defined in a range $[0, 3]$ with a distribution function $f(x) = x^2$.

   a. Determine the *pdf*, $p(x)$, and *cdf*, $P(x)$, of this random variable.

   b. Write a program for selecting x using a random number $\eta \in [0, 1]$. Using 1,000 and 50,000 samples, calculate the mean of the distribution. Compare this to the theoretical result.

5. Write computer programs to sample from the following distribution functions. Plot a histogram of the sampled x values for 100,000 samples. Compare this to the *pdf*.

   a. $f(x) = 1 + x + x^3$, $x \in [0, 1]$ (use probability mixing)

   b. $f(x) = 1 + x - x^3$, $x \in [0, 1]$ (use probability mixing)

   c. $f(x) = e^{-x}$, $x \in [0, \infty)$ (use analytic FFMC)

6. Using the rejection technique, estimate the area inside a circle of radius 1 but outside of a square inscribed inside the circle. This is shown as the gray area in Figure 2.10. Compare this calculated area to the "true" area.

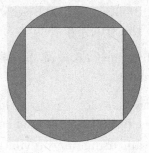

Figure 2.10 Problem 6

7. Consider two circles with radii R_1 and R_2 and distance d between their origins, as shown in Figure 2.11. Write a Monte Carlo algorithm to determine the area of the overlapped region for:

   a. $R_1 = d = 1$,   and $R_2 = 0.5$

   b. $R_1 = R_2 = 1$,   and $d = 1.5$

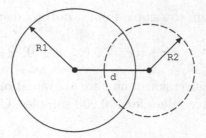

Figure 2.11   Problem 7

8. Repeat Problem 7 for two spheres.

9. Write a Monte Carlo algorithm to determine $\bar{x}$ and σ_x in the square depicted in Figure 2.12.

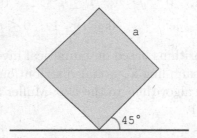

Figure 2.12   Problem 9

10. Write a Monte Carlo algorithm to determine the area of the section (shaded) as depicted in Figure 2.13.

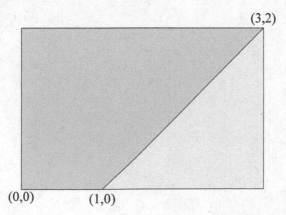

Figure 2.13   Problem 10

11. Write a program to sample from a normal distribution given by:

$$f(x) = e^{-\frac{x^2}{2}}, \qquad x \in [0, \infty)$$

Use numerical inversion or numerical evaluation. Plot a histogram of the sampled x values for 100,000 samples. Compare this to the *pdf*.

12. For the following multidimensional density functions, derive the FFMC for each variable.

   a. $f(x, y) = x^2 + y^2, \quad 0 \le x \le 1, \quad 0 \le y \le 1$

   b. $f(x, y, z) = xyz, \quad 0 \le x \le 1, \quad 0 \le y \le 1, \quad 0 \le z \le 1$

   c. $f(x, y, z, w) = x + y^2 + z + w^3,$
      $0 \le x \le 1, \quad 0 \le y \le 1, \quad 0 \le z \le 1, \quad 0 \le w \le 1$

   d. $f(x, y) = xye^{x^2 + y^2}, \quad 0 \le x \le 1, \quad 0 \le y \le 1$

13. Write two algorithms based on numerical inversion and rejection technique for sampling a normal distribution. Compare the efficiency of these algorithms to the Box-Muller algorithm discussed in Section 2.7.1.

14. Write two algorithms based on numerical inversion and rejection technique for sampling the Watt spectrum. Compare the efficiency of these algorithms to that discussed in Section 2.7.2.

   Write an algorithm for sampling $sin(x)$. Compare the efficiency of your algorithm to the von Neumann formulation given in Section 2.7.3.

CHAPTER 3

Random Number Generator (RNG)

CONTENTS

3.1 INTRODUCTION

Random numbers (RNs) are an essential part of any Monte Carlo simulation. The quality of any Monte Carlo simulation depends on the quality (or randomness) of the random numbers used. A high degree of randomness is achieved if the random numbers follow a uniform distribution. Therefore, we need to devise approaches that yield sequences of numbers that are random, have a long period before repeating, and do not require significant resources to generate.

Early implementation of random number generators on computers can be traced back to John von Neumann who used them for Monte Carlo simulation related to the Manhattan project (1941–1945). Since then, numerous approaches for the generation of random numbers have been developed [15, 40, 59, 62, 63, 64, 70, 69, 58]. As indicated by L'Ecuyer, the average computer user usually thinks that the problem of generating uniform random numbers by computer has been solved. Although there has been significant progress in the development of "good" generators, there are still "bad" or "unsuitable" generators, which yield undesirable results. Hence, any Monte Carlo user should be well aware of the issues and limitations.

For completeness, in this chapter, both experimental and algorithmic techniques are introduced and a selection of randomness tests are discussed. Through examples, it is demonstrated that a random number generator's randomness and period are highly dependent on "correct" selection of several parameters. These examples demonstrate that minor changes in parameters can have a significant impact on the estimated random numbers.

In this chapter, we will review experimental and algorithmic approaches used for the determination of random numbers, examine the behavior of the random number generators, review several randomness tests, and elaborate on the impact of the parameters on the length of sequence of random numbers and randomness of generated random numbers.

3.2   RANDOM NUMBER GENERATION APPROACHES

There are two common approaches for generating random numbers: 1) Experimental; 2) Algorithmic.

1. Experimental (look-up tables, online): An experiment (physical process) is used to generate a sequence of random numbers that are saved in computer memory as tables. Examples are: (a) flip a coin or toss a die; (b) draw balls from an urn, i.e., lottery; (c) spin the marked disc introduced in Chapter 2; and, (d) measure the position of a dart from the center in a dart game. In an approach by Frigerio and Clark [36], the number of disintegrations of radioactive material over a given time interval,

e.g., $20ms$ is counted and, if the number is odd, a 0-bit is recorded, otherwise a 1-bit is recorded. Then 31-bit numbers are formed. This process produces $< 6,000$ numbers over one hour. Obviously, it is quite slow to be used during a Monte Carlo simulation. (A tape containing 2.5×10^6 random numbers is available at the Argonne National Lab's (ANL) code center.) Another technique is to monitor the components, e.g., memory of a computer to generate random numbers. Note that this is done while performing the actual simulation.

2. Algorithmic (or deterministic): An algorithm is used to generate random numbers. This approach, because of its deterministic nature, is commonly referred to as pseudo random number generator (PRNG)and its associated numbers are referred to as pseudo random numbers (PRNs).

Each approach has its advantages/disadvantages, which directly impact its use. The important factors for selecting the "right" approach include:

1. **Randomness:** Random numbers should be truly random; i.e., they should assume a uniform distribution. In the experimental approach, randomness is achieved if the procedure follows a uniform distribution. In the algorithmic approach, the generated sequence has to satisfy several statistical tests.

2. **Reproducibility:** To test a simulation algorithm or to perform sensitivity/perturbation studies, it is necessary to be able to reproduce the random number sequence.

3. **Length of the sequence of random numbers:** To perform a Monte Carlo simulation for realistic engineering problems, millions of random numbers are needed; hence, the generator has to be able to produce a large number of random numbers.

4. **Computer memory:** The generator should not consume too much computer memory. (Note that this issue may not be important for future generation computers.)

5. **Generation time:** The amount of time (engineer/analyst) that it takes to generate a sequence of random numbers should not be significant, e.g., days/months.

Table 3.1 Comparison of experimental and algorithmic random number generators

Factor	Experimental		Algorithmic
	Table	online	
Randomness	good	good	to be tested
Reproducibility	yes	no	yes
Period[1]	limited	n/a	limited
Computer memory	large	small	small
Generation time[2]	long	n/a	n/a
Computer time	short	long	short

[1] Maximum length of the sequence of RNs.
[2] Time needed to generate RNs.

6. **Computer time:** The amount of computer time needed to generate the numbers should be significantly shorter than the actual simulation.

Table 3.1 compares the experimental and algorithmic generators based on the aforementioned factors.

In practice, the algorithmic approach is the preferred choice mainly because its sequence is reproducible and it requires minimal effort (computer resource, engineer time). In using any algorithmic generator, it is necessary to perform a series of randomness tests and to determine/measure the period at which the generator repeats its sequence, i.e., loss of randomness.

The following sections discuss different algorithmic random (pseudo random) number generators and related randomness testing approaches.

3.3 PSEUDO RANDOM NUMBER GENERATORS (PRNGS)

A good PRNG has to yield a sequence of uniformly distributed random numbers in a range $(0, 1)$. Further, it should have a long period and must pass a series of randomness tests. Here, we will introduce commonly used generators including: (a) congruential; and, (b) multiple recursive.

3.3.1 Congruential Generators

A congruential generator is an integer generator proposed by D. H. Lehmer [64]. It uses the following formulation:

$$x_{k+1} = (ax_k + b), \quad mod \ M, \ for \ b < M \tag{3.1}$$

where $x_0(seed)$, a, b, and M are given integers. M is the largest integer represented by a computer; e.g., on a binary machine with a 32-bit word length, the largest unsigned integer is $2^{32} - 1$ and the largest signed integer is $2^{31} - 1$. The modulus function determines the remainder of $(\alpha = ax_k + b)$ divided by M, e.g., (35) mod 16 is equal to 3.(Note that the term congruential stems from the fact that the ratios of $\frac{\alpha}{M}$ and $\frac{x_{k+1}}{M}$ have the same remainder, i.e., x_{k+1} is congruent to α modulo M.) It is worth noting that Equation 3.1 is referred to as a *linear congruential* generator, and, if $b = 0$, it is called a *multiplicative congruential* generator.

The methods discussed here will give a random integer x in the range $[0, M{-}1]$. To convert the random integer generated here into a random number in the range of $[0, 1)$, we use the relationship $\eta = \frac{x}{M-1}$.

It is instructive to use a simple example to discuss specifics of the congruential generators and related issues. Let's consider a linear congruential generator [20] given by

$$x_{k+1} = (5x_k + 1), \quad mod \ 16 \tag{3.2}$$

Considering a *seed* equal to 1, the sequence of random numbers can be presented on a *random number cycle* (clockwise) as shown in Figure 3.1.

The cycle in Figure 3.1 exhibits the expected behavior of a generator as follows:

1. The sequence has a period of 16; i.e., equivalent to the modulus, $M = 2^4$, and the largest number is $2^4 - 1$.

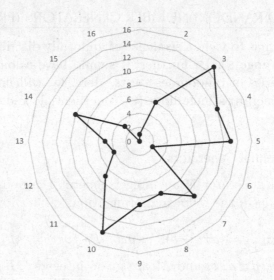

Figure 3.1 Schematic of RN cycle for a generator given by Equation 3.2. Note that the contours give the RN values, and the radial indices refer to order of the RN in the sequence.

2. Selection of any other number ($< M = 16$) as seed results in a cyclical shift of the above sequence.

It is instructive to examine the effect of multiplier (a) on the period of the above generator. Table 3.2 presents the period of the generator for different multipliers between 3 and 15.

Table 3.2 indicates that if the multiplier (a) is equal to 5, 9, or 13, a full period of 16 is obtained, otherwise, partial periods of 2, 4, or

Table 3.2 Impact of *multiplier*(a) on the *period* of the PRNG (Eq. 3.2)

Parameter	Value												
Multiplier	3	4	5	6	7	8	9	10	11	12	13	14	15
Period	8	2	16	4	4	2	16	4	8	2	16	4	2

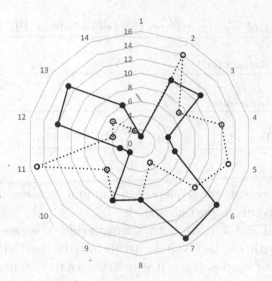

Figure 3.2 Random number sequences for multipliers *9 (solid) and 13 (dotted)*

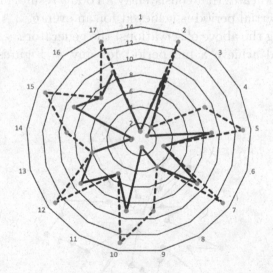

Figure 3.3 Random number sequences for multipliers 8 (solid) and 12 (dotted)

8 are achieved. More specifically, Figure 3.2 presents the sequences of random numbers for multipliers 9 (solid) and 13 (dotted).

Table 3.3 Impact of *constant*(*b*) on the *period* of the PRNG (Eq. 3.2)

Parameter	Value						
Constant	2	3	4	5	6	7	8
Period	8	16	2	16	8	16	4

Figure 3.3 presents the sequences of random numbers for multipliers 8 and 12 that result in partial periods.

As expected in the cases with a full period, the random numbers form shapes with distinct values, while all the partial period cases exhibit repeating shapes. Next, it is instructive to examine the effect of the constant (*b*).For the case with multiplier 5, we determine the effect of constant (*b*) on the period of the generator. Table 3.3 compares the period of generator for different *b* values in a range of 2 to 8.

Table 3.3 indicates that consistently an odd *b* results in a full period, while only a partial period is achieved for an even *b*.

Considering the above observations, the generator $x_{k+1} = (5x_k + 1)$ *mod* 2^n should achieve a full period for any n. Figures 3.4 and 3.5

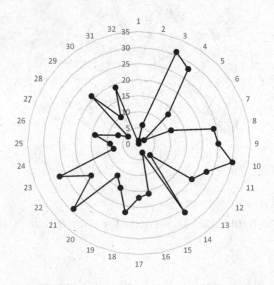

Figure 3.4 Random number sequence the Eq. 3.2 PRNG with $M = 2^5$

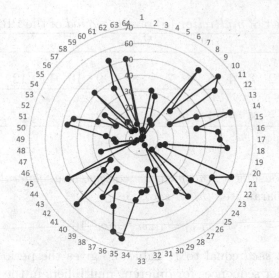

Figure 3.5 Random number sequence the Eq. 3.2 PRNG with $M = 2^6$

show the sequences of random numbers for n equal to 5 and 6, respectively.

Figures 3.4 and 3.5 indicate that, as expected, 32 and 64 random numbers, i.e., full period are generated corresponding to M equal to 32 and 64, respectively.

To achieve a full period for a linear congruential generator, we can use the *corollary* to the theorem by Hull and Dobell [53] which sets the properties of different parameters including multiplier, constant and modulus. Table 3.4 presents these properties.

Table 3.4 Properties of the parameters in a linear congruential generator for achieving a full period

Parameter	Value	Comment
multiplier (a)	4N+1	$N > 0$
contact (b)	odd number	
Modulus (M)	2^k	$k > 1$

Table 3.5 Impact of *multiplier* (*a*) on the *period* of the PRNG (Eq. 3.3)

Parameter	Value													
Multiplier	2	3	4	5	6	7	8	9	10	11	12	13	14	15
Period	-	4	-	4	-	2	-	2	-	4	-	4	-	2

Now, we consider a *multiplicative congruential generator* by setting the constant parameter *(b) to zero*, i.e.,

$$x_{k+1} = (ax_k) \; mod \; 16 \tag{3.3}$$

Considering a *seed* equal to 1, Table 3.5 gives the period of different random number sequences for different multipliers in the range of 2 to 15.

Figure 3.6 shows random number sequences generated by the multiplicative generator, Equation 3.3, for two multipliers including 3 and 11. This demonstrates the partial period of the generator as each case leads to polygons that are terminated by repeating the seed. Similar

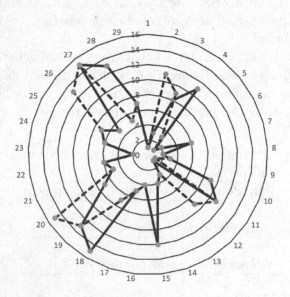

Figure 3.6 Random number sequence for multipliers 3 (solid) and 11 (dotted), Eq. 3.3 PRNG

results are obtained for 5 and 13 multipliers. These results indicate that for cases with $a = 8N + 3$ or $a = 8N + 5$ for $N \geq 0$, a multiplicative congruential generator yields a period of 2^{k-2}, where k corresponds to the exponent of modulus, e.g., $16 = 2^4$. Now, considering a realistic generator with an odd multiplier (16339) and an even mod, 2^k, we will determine its period for any power N using the algorithm given in Table 3.6.

Note that the mod function can be replaced by the following formulation

$$ixx(i) = iab - INT(iab/mod) * imod \qquad (3.4)$$

Note that the implementation of this algorithm is dependent on knowledge of integer operations of a specific system. (For a discussion on this, see Appendix 1.)

If we consider different powers k for modulus (mod) of 2^k, we can demonstrate that the period of each sequence is 25% of the modulus.

Table 3.6 An algorithm for a linear congruential RNG

Algorithm	Comment
$read*, x_0, multi, const, imod$	*read seed, multiplier, constant, and modulus*
$ixx(1) = x_0$	*initialize the RN array*
$do\ i = 2, imod$	*do-loop to generate RNs*
$iab = multi * ixx(i - 1) + const$	*calculate $ax_{i-1} + b$*
$ixx(i) = mod(iab, imod)$	*using 'mod' function*
	obtain a new RN
$if(ixx(i).lt.0)then$	*check for positivity of RN*
$ixx(i) = ixx(i) + imod$	
$endif$	
$if(ixx(i).eq.x_0)then$	*Determine the period*
$iperiod = i - 1$	
$goto\ 10$	
$else$	
$iperiod = i$	
$endif$	
$enddo$	
$10\ continue$	

To achieve a large period ($>$ 25% of modulus) for a multiplicative congruential generator, it is demonstrated [20] that a period of M–1 can be achieved if we consider the following:

1. M is a prime number

2. a multiplier is a primitive of M

Park and Miller [79] demonstrated that for a modulus of 2^{31}–1 and a multiplier of 16807, a period of $M - 1$ is obtained. For further information on the best random number generators, the reader should consult Bratley, Fox, and Schrage [15], Park and Miller [79], Zeeb and Burns [121], L'Ecuyer [63], Gentle [39], and Marsaglia [69].

In summary, a congruential generator with a reasonable period is satisfactory for most physical simulations, because the physical system introduces randomness by applying the same random numbers to different physical processes. This statement, however, is not always true for the solution of a mathematical problem.

3.3.2   Multiple Recursive Generator

A group of PRNGs referred to as multiple recursive generators [39] can be expressed by

$$x_{k+1} = (a_0 x_k + a_1 x_{k-1} + ... + a_j x_{k-j} + b), \ mod \ M \qquad (3.5)$$

One initiates the generator by selecting j+1 random numbers (possibly from a simpler generator). The length and the randomness (or statistical properties) of the generator depend on the values of a_j, b, and M. A special case of this type of generator is the Fibonacci generator.

A Fibonacci generator is a floating point generator. It is characterized by computing a new number by a combination (difference, sum, or product) of two preceding numbers. For example, the formulation [55]:

$$x_k = x_{k-17} - x_{k-5}, \qquad (3.6)$$

is a Fibonacci generator of lags 17 and 5. The generated sequence depends on the choice of the initial $x_k's$, e.g., 17. Since the above Fibonacci formulation x_{k+1} depends upon 17 preceding values, its period (p) can be large, i.e., $p = (2^{17}-1)2^n$, where n is the number of bits in the fractional part of x_i, the mantissa. For example, for 32-bit

Table 3.7 A list of recommended Fibonacci generators

Generator	Expected period
$x_k = x_{k-17} - x_{k-5}$	$(2^{17} - 1) * 2^{24} = 2.2 * 10^{12}$
$x_k = x_{k-31} - x_{k-13}$	$(2^{31} - 1) * 2^{24} = 3.6 * 10^{16}$
$x_k = x_{k-97} - x_{k-33}$	$(2^{97} - 1) * 2^{24} = 2.7 * 10^{36}$

floating point arithmetic, $n = 24$; hence, p is $\sim 2^{41}$ or 10^{12}. Because of the large expected period, a Fibonacci generator is a good choice for some large problems. For even larger problems on supercomputers, Fibonacci generators with lags 97 and 33 have been used.

To start a Fibonacci generator, it is necessary to generate the initial random numbers, e.g., 17 numbers. One approach is to represent each initial random number in its binary form

$$r = \frac{r_1}{2} + \frac{r_2}{2^2} + \text{........} + \frac{r_m}{2^m}, \quad for \; m \leq n(mantissa) \quad (3.7)$$

where r_i is a binary number, i.e., 0 or 1. A simpler congruential generator is needed to set each bit, r_i as 0 or 1. For example, we may set r_i to either 0 or 1 depending on whether the output of the integer congruential generator is greater or less than zero. Hence, the quality of a Fibonacci formulation does depend on the quality of the initial numbers, or the simpler integer generator. A list of recommended Fibonacci generators is given in Table 3.7.

Note that the largest period that a congruential generator can have when using 32-bit integers is 2^{32}, or $4.3 * 10^9$, which is significantly smaller (at least by a factor of ~ 500) than any one of the Table 3.7 recommendations.

3.4 TESTING RANDOMNESS

A pseudo random number generator is acceptable if it passes a series of randomness tests. There are numerous randomness tests [59, 68], which look at various parameters for examining independence and uniformity of a random number sequence. In the remainder of this section, we will discuss a set of tests that examine the random numbers based on their digits and as whole numbers [77].

3.4.1 $\chi^2 - Test$

The $\chi^2 - test$ measures the deviation between the sample and the assumed probability distribution, i.e., hypothesis. The formulation of the $\chi^2 - test$ is given by

$$\chi^2 = \sum_{i=1}^{n} \frac{(N_i - Np_i)^2}{Np_i} \tag{3.8}$$

where $\{p_1, p_2,..., p_n\}$ is a set of hypothetical probabilities associated with N events falling into n categories with observed frequencies of N_1, $N_2,...,N_n$. Note that this test examines the whole sampled distribution at once in comparison to a hypothetical distribution, and is, in this sense, more general than the examination of a sample mean, sample variance, etc. For large values of N, the random variable χ^2 approximately follows the χ^2- *distribution density function* with $n-1$ degrees of freedom.

3.4.1.1 $\chi^2 - distribution$

A random variable $w = \chi^2$ has a $\chi^2 - distribution$ [19], if it follows a probability density function given by

$$f_m(w)dw = w^{\frac{m}{2}-1} 2^{\frac{m}{2}} \Gamma(\frac{m}{2}) e^{-\frac{w}{2}} dw \tag{3.9}$$

where m is a positive integer, referred to as the number of degrees of freedom, Γ is the gamma function, and $w > 0$. Generally, one is interested in finding the probability of getting a $w = \chi^2$ that is less than a given value of χ_0^2; however, the available tables for the $\chi^2-distribution$, commonly, give probabilities for $\chi^2 \geq \chi_0^2$. Hence, it is necessary to use the complement of the probabilities, i.e.,

$$P(\chi^2 \leq \chi_0^2) = 1 - P(\chi^2 \geq \chi_0^2) \tag{3.10}$$

where

$$P(\chi^2 \geq \chi_0^2) = \int_{\chi_0^2}^{\infty} dw f_m(w) \tag{3.11}$$

It is worth noting that the $\chi^2 - distribution$ approaches a normal distribution as m increases, and the mean and the variance of the distribution are equal to m and $2m$, respectively.

Table 3.8 An example for χ^2 table ($P(\chi^2 \geq \chi_0^2)$)

Degrees of freedom	0.99	0.95	0.05	0.01	0.001
1	0	0.004	3.84	6.64	10.83
2	0.02	0.103	5.99	9.21	13.82
3	0.115	0.352	7.82	11.35	16.27
4	0.297	0.711	9.49	13.28	18.47
5	0.554	1.145	11.07	15.09	20.52
6	0.872	1.635	12.59	16.81	22.46
7	1.239	2.167	14.07	18.48	24.32
8	1.646	2.733	15.51	20.09	26.13
9	2.088	3.325	16.92	21.67	27.88
10	2.558	3.94	18.31	23.21	29.59

3.4.1.2 Procedure for the use of χ^2 – test

The chi-squared (χ^2) values are obtained according to Equation 3.8, and then these values are compared with the determined values given in χ^2 tables, such as Table 3.8.

Commonly, we compare the estimated values, Equation 3.8, to predicted χ_0^2 values that correspond to 5% and 95% probabilities. As an example, divide the $(0, 1)$ space into 10 equally spaced bins, i.e., 0.0–0.1, 0.1–0.2, etc. For an RNG, we expect that every bin has an equal probability of 0.1, i.e., $p_i = 0.1$. After taking N random samples, we get N_i results in each bin. If we calculate χ^2, as in Equation (3.8), it is expected that the estimated values of χ^2 with 9 degrees of freedom will be between the predicted values of 3.325 and 16.919 at 5% and 95% probabilities, respectively. If χ^2 is smaller than 3.325, the RNG fails the test because it provides values that are closer than the "truth," i.e., hypothesis predicts. Conversely, if the χ^2 value is larger than 16.919, it means the RNG-generated numbers are too "variant," beyond the expected level.

3.4.2 Frequency test

In this test, one counts the number of occurrences of every digit (0 through 9) by parsing digits in every random number generated by a

RNG. Knowing the expected occurrence of $1/10$, one can calculate χ^2 values to determine the degree of randomness of the generator.

3.4.3 Serial test

The serial test is an extension of the frequency test to pairs of digits. For any selected digit, one counts the occurrence of every other digit following the selected digit. Again, the expected chance of occurrence is $1/10$, and χ^2 values provide information on the degree of randomness of the generator.

3.4.4 Gap test

In the gap test, a particular digit, e.g., 0, is selected and the frequency of occurrence of nonzero digits between successive $0's$ is determined. For a single gap, the expected frequency is $9/100$. Again, the χ^2 values are calculated to measure the quality of generator.

3.4.5 Poker test

In this test, the digits are partitioned into groups of five and the relative frequencies of "five of a kind," "four of a kind," etc., are determined. Again, χ^2 values are calculated to measure the quality of the generator.

3.4.6 Moment test

Given k^{th} moment of variable y is defined by

$$\mu_k = \int_a^b dy y^k p(y) \tag{3.12}$$

Since a random number sequence should be distributed uniformly within the unit interval, $[0, 1]$, then, for a random variable η with probability density $p(\eta) = 1$, the kth moment of η is given by

$$\int_0^1 d\eta \eta^k = \frac{1}{k+1} \tag{3.13}$$

Consequently, the randomness of a sequence of random numbers, $\eta's$ can be tested by examining the following equality

$$\sum_{i=1}^N \eta_i^k \simeq \frac{1}{k+1} . \tag{3.14}$$

Note that for $k = 1$, the above equation reduces to a simple average of random numbers that is equal to 0.5.

3.4.7 Serial correlation test

The serial correlation coefficient for a sequence of random numbers x_i of length N and lag j is given by

$$\rho_{N,j} = \frac{\frac{1}{N}\sum_{i=1}^{N} x_i x_{i+j} - [\frac{1}{N}\sum_{i=1}^{N} x_i]^2}{\frac{1}{N-1}\sum_{i=1}^{N} x_i^2 - [\frac{1}{N}\sum_{i=1}^{N} x_i]^2} \tag{3.15}$$

If x_i and x_{i+j} are independent and N is large, then the correlation coefficient $(\rho_{N,j})$ should follow a normal distribution with mean $(-\frac{1}{N})$, and standard deviation $(\frac{1}{\sqrt{N}})$. In addition to comparing with a normal distribution, we may use the χ^2 values to measure the degree of randomness.

3.4.8 Serial test via plotting

As random numbers are generated, they are combined, e.g., paired or 2-tuple, (x_1, x_2), (x_3, x_4),..., and each pair (or combination) is plotted as a point in a unit square. If there is any obvious pattern, e.g., striped, then it can be concluded that the numbers are serially correlated.

3.5 EXAMPLE FOR TESTING A PRNG

For a linear congruential generator given by

$$x_{k+1} = (ax_k + b) \; mod \; M \tag{3.16}$$

Our objective is to examine the effect of seed (x_0), multiplier (a), and constant (b) on the randomness of the generator. We compare different sets of parameters by determining the period of the sequence, average, i.e., first moment of the random numbers, and plotting of positions within a $3 - D$ domain.

3.5.1 Evaluation of PRNG based on period and average

First, we examine the effect of variations of the *seed* in a multiplicative congruential generator with an odd multiplier (65539), as given in Table 3.9.

Table 3.9 Impact of *seed* on a multiplicative congruential generator

Parameter	case	seed	a	b	M	period	$\frac{period}{M}$	average
seed	1	1	65539	0	2^{24}	4194303	25%	0.500
	2	69069	65539	0	2^{24}	4194303	25%	0.500
	3	16806	65539	0	2^{24}	2097152	12.5%	0.500
	4	1024	65539	0	2^{24}	4096	0.024%	0.500
	5	4096	65539	0	2^{24}	1024	0.006%	0.500

Table 3.9 indicates that the seed has a major impact on the period of the sequence. It confirms our earlier results that, for an odd multiplier with an even modulus, a 25% period is achieved. In Case 3, with an even seed, but not a multiple of 2, a reduced period of 12.5% is observed, while for Cases 4 and 5, with power-2 seeds, poor periods are expected. Although, even in these cases, the average is correct. For a better understanding of the impact of the seed, in Table 3.10, we determine the distribution of random numbers within the range of $[0, 1]$.

Table 3.10 demonstrates that Cases 4 and 5 not only have a very short period, but also result in a biased distribution of random numbers.

Second, we examine the effect of variations of the multiplier (a) for a fixed odd seed of 1. Table 3.11 compares the periods and averages

Table 3.10 Impact of *multiplier* on a multiplicative congruential generator

Interval	Case 1	Case 2	Case 3	Case 4	Case 5
0.0-0.1	10%	10%	10%	10.01%	10.06%
0.1-0.2	10%	10%	10%	10.01%	10.06%
0.2-0.3	10%	10%	10%	10.01%	9.96%
0.3-0.4	10%	10%	10%	9.99%	9.96%
0.4-0.5	10%	10%	10%	9.99%	9.96%
0.5-0.6	10%	10%	10%	10.01%	10.06%
0.6-0.7	10%	10%	10%	10.01%	10.06%
0.7-0.8	10%	10%	10%	10.01%	9.96%
0.8-0.9	10%	10%	10%	9.99%	9.96%
0.9-1.0	10%	10%	10%	9.99%	9.96%

Table 3.11 Impact of *multiplier* on a multiplicative congruential generator

Parameter	case	*seed*	*a*	*b*	*M*	*period*	$\frac{period}{M}$	average
multiplier	6	1	69069	0	2^{24}	4194303	25%	0.500
	7	1	1024	0	2^{24}	3	0	0.021
	8	1	1812433253	0	2^{24}	4194303	25%	0.500
	9	1	4096	0	2^{24}	2	0	0.0
	10	1	1	0	2^{24}	1	0	0.0

Table 3.12 Impact of *constant* on a linear congruential generator

Parameter	case	*seed*	*a*	*b*	*M*	*period*	$\frac{period}{M}$	average
constant	11	1	65539	1	2^{24}	4194304	25%	0.500
	12	1	65539	1024	2^{24}	4194304	25%	0.500
	13	1	65539	4096	2^{24}	4194303	25%	0.500
	14	1	65539	65539	2^{24}	8388608	50%	0.500
	15	1	65539	69069	2^{24}	8388608	50%	0.500
	16	1	65539	$2^{24} - 1$	2^{24}	8388608	50%	0.500

for odd and even multipliers. Table 3.11 indicates that the multiplier has significant impact on the performance of a generator. Basically, if the multiplier is a power of 2 (similar to the modulus) (Cases 7 and 9), the generator fails; otherwise, for a large odd multiplier, the generator behaves as expected. Third, we examine the effect of constant (*b*), for a fixed seed and multiplier, as given in Table 3.12. It is evident that the impact of constant (*b*) is not as significant as either multiplier or seed. However, the results indicate that the constant can significantly impact the period for a given choice of multiplier and seed. To explore this further, in Table 3.13, we examine small changes in the multiplier from 65539 to 65549 with an increment of 2, i.e., only odd multipliers, and fix seeds and constants of 1. The above data Table 3.13 indicates that the generator has truly significant dependency on the multiplier because Cases 18, 20, and 22 show full periods as compared to the rest with partial periods. The cases with full periods have a similar characteristic; their multiplies can be represented by the following formulations:

Table 3.13 Impact of small changes of *multiplier* on a linear congruential generator

Parameter	case	seed	a	b	M	period	$\frac{period}{M}$	average
multiplier	17	1	65539	1	2^{24}	4194304	25%	0.500
	18	1	65541	1	2^{24}	16777216	100%	0.500
	19	1	65543	1	2^{24}	4194303	25%	0.500
	20	1	65545	1	2^{24}	16777216	100%	0.500
	21	1	65547	1	2^{24}	8388608	50%	0.500
	22	1	65549	1	2^{24}	16777216	100%	0.500

Case 18: $multiplier = 65539 = 4 \times 16385 + 1$

Case 20: $multiplier = 65541 = 4 \times 16385 + 3$

Case 22: $multiplier = 65543 = 4 \times 16385 + 5$

The above observation for Case 18 provides the necessary condition for achieving a full period, as the Case 18's properties are similar those stated in the corollary to the Hull and Dobell's theory (Table 3.4).

3.5.2 Serial test via plotting

Here, we examine two linear generators with different multipliers. Our aim is to demonstrate the potential benefit of the plotting approach, which can exhibit correlations among the random numbers.

First, we examine, two generator cases from Table as follows:

Case 17: $a = 65539$, $seed = 1$, $b = 1$, $mod = 2^{24}$

Case 18: $a = 65541$, $seed = 1$, $b = 1$, $mod = 2^{24}$

As indicated, Case 17 led to a period of 25% of the modulus and Case 18 resulted in a period equal to the modulus. To examine the randomness via plotting, we generated 3-tuple positions (x, y, z) using every three consecutive random numbers and then marked the positions in a 3-D domain. Figures 3.7 and 3.8 correspond to Cases 17 and 18, respectively.These figures indicate that both generators lead to correlated (especially, Case 17) sets of random numbers irrespective of achieving a full period (Table 3.13).

Finally, we examine Case 23 with a multiplier of 16333, while other parameters remain the same as Cases 17 and 18. This multiplier results in a period equal to the modulus, and Figure 3.9 shows its 3-tuple diagram with no visible correlation.

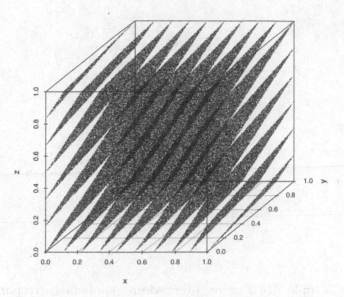

Figure 3.7 3-tuple distribution of random numbers corresponding to Case 17

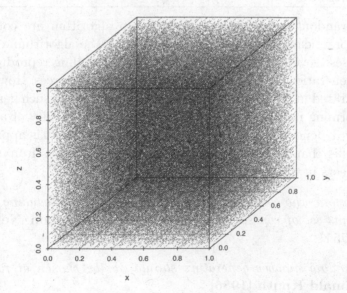

Figure 3.8 3-tuple distribution of random numbers corresponding to Case 18

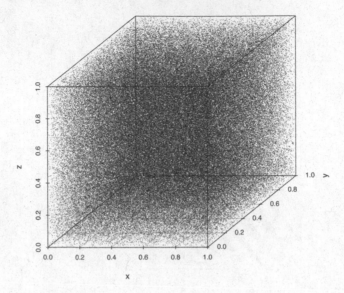

Figure 3.9 3-tuple distribution of random numbers corresponding to Case 23

3.6 REMARKS

Pseudo random numbers generated via an algorithm are commonly used in practical applications. This is because the algorithmic generators possess several favorable characteristics including reproducibility, ease of generation, and minimal use of computer resources. However, as demonstrated in this chapter, it is essential to examine such generators by performing numerous randomness tests, as the quality of a PRNG is highly dependent on using the "right" parameters. It is appropriate to end this chapter with the following profound quotes from some of the pioneers in this field:

> *Anyone who considers arithmetical methods of producing random digits is, of course, in a state of sin.*– **John von Neumann [1951]**

> *Random number generators should not be chosen at random.*– **Donald Knuth [1986]**

PROBLEMS

1. Demonstrate that the largest possible period for a congruential generator is equal to the modulus, and that the largest period for a Fibonacci generator of lags l and k (with $l > k$) is equal to $(2^l - 1) \cdot 2^N$, where N is equal to the mantissa of a real computer word.

2. Write a program to demonstrate that for a linear congruential generator with modulus $(m = 2N)$, constant (b) equal 1, in order to achieve a full period, its multiplier (a) has to be equal to $4K + 1$. Consider $N = 5$, 6, and 7; and $K = 2$ and 9.

3. Determine the length of periods of the last three generators in Table 3.14, and prepare 2-tuple and 3-tuple diagrams for each generator.

4. Write a program for a linear congruential RNG given by:

$$x_{k+1} = (ax_k + b) \ mod \ M$$

Measure the quality of the RNG for the combinations of different parameters a, b, and x_0 in Table 3.14.

Use modulus $M = 2^{31}$. Determine the period of each generator, and perform "*frequency*," "*moment*," and serial testing via plotting tests. Tabulate, plot, and discuss your results. For the frequency test, select five digits to test.

5. One of the pioneers in the area of Monte Carlo methods, John von Neumann, while working on the Manhattan project, proposed the

Table 3.14 Combinations of parameters

Case	a	b	x_0
(a)	1812433253	1	69069
(b)	1812433253	69069	69069
(c)	69069	69069	69069
(d)	65539	0	69069
(e)	65539	0	1024
(f)	1024	0	69069

Table 3.15 Congruential random number generators[*]

a	b	M	Author
23	0	$10^8 + 1$	Lehmer
$2^7 + 1$	1	2^{35}	Rotenberg
7^5	0	$2^{31} - 1$	GGL
131	0	2^{35}	Neave
16333	25887	2^{15}	Oakenfull
3432	6789	9973	Oakenfull
171	0	30269	Wichmann-Hill

[*] Source: Rade, L. and B. Westergren. 1990, BETA mathematics handbook, 2nd ed. (With permission)

following random number generator:

$$x_{k+1} = middle\,digits(x_k * x_k)$$

Examine the quality of this generator by preparing 2-tuple and 3-tuple diagrams, and determine the period for different seeds. Use 4-digit numbers for the x_k, and the middle digits function extracts the middle 4 digits of the resulting 8-digit number (padding zeros are added to x_{k+1} if it is less than 8 digits).

6. Examine the quality of the different congruential generators given in Table 3.15 [83] by preparing 2-tuple diagrams and determining the period of each generator for a given seed.

7. Write a program to estimate the mean free path (*mfp*) for a particle traveling in a medium, where the path length is sampled from a probability density function given by $p(r) = \Sigma e^{-\Sigma r}$, and $\Sigma = 1.00\ cm^{-1}$. For checking the solution convergence, consider the relative difference of 0.1% between the simulation result and the theoretical prediction.

For this simulation, consider three RNG cases discussed in Section 3.5.2.

8. Show that the largest integer number on a computer with a word length of 32 bits, where one bit is used for sign, is equal to $2^{31} - 1$.

CHAPTER 4

Fundamentals of Probability and Statistics

CONTENTS

4.1   INTRODUCTION

This chapter is devoted to a discussion of a set of concepts and formulations in probability and statistics that, in the author's opinion, are important for performing and understanding the results of Monte Carlo simulations. The reader should consult standard books on probability and statistics for a more in-depth understanding of the content presented in this chapter.

Statistical procedures are needed in dealing with any random process. These procedures provide a means of describing and indicating trends or expectations (averages), which often have an associated degree of reliability (standard deviation, confidence level). In short, statistics make use of scientific methods of sampling, i.e., collecting, analyzing, and interpreting data when the population (probability density function) is unknown.

The theory of statistics rests on the theory of probability. The latter often furnishes much of the basics or underlying structure of the former. Probability theory deals with determining the likelihood that an unknown sample turns out to have stated characteristics; here, it is assumed that the population is known. In contrast, the theory of statistics deals with sampling an unknown population in order to estimate its composition, i.e., population or probability density [19, 39, 83]. As discussed in earlier chapters, the Monte Carlo approach is a statistical method that utilizes random numbers to sample an unknown population, e.g., particle history and consequently, evaluates the expected outcome of a physical process.

To present the results of a statistical analysis/simulation, commonly one attempts to estimate the following three quantities:

1. The sample average

2. The sample variance/standard deviation and relative uncertainty

3. The confidence level

This chapter will discuss the fundamentals of probability and its use for the derivation of statistical formulations, such as sample average and sample variance. The concepts of precision and accuracy are discussed as well. A number of standard probability density functions are introduced and their relations are examined. Two important limit theorems(de Moivre–Laplace and Central) for examining the precision of sample averages are introduced and their uses are discussed. These theorems are based on large sample sizes and convergence to a normal distribution; however, it should be noted that we are commonly dealing with finite sample sizes, and, therefore, we need to introduce and elaborate on methodologies for testing the "normality" of the results, such as the Student t-distribution.

4.2 EXPECTATION VALUE

4.2.1 Single variable

Given a continuous random variable x that has a probability density function (pdf), $p(x)$, in a range of $[a, b]$, then the expectation value (or true mean) of function $g(x)$ is given by

$$E\left[g(x)\right] = \frac{\int_a^b dx p(x) g(x)}{\int_a^b dx p(x)} \tag{4.1}$$

Since the pdf is a normalized function, then the denominator of the above equation is equal to 1. Hence, the expectation formulation reduces to

$$E\left[g(x)\right] = \int_a^b dx p(x) g(x) \tag{4.2}$$

For a discrete random variable, x_i, of N outcomes, the expectation value formulation reduces to

$$E\left[g(x_i)\right] = \sum_{i=1}^N p(x_i) g(x_i) \tag{4.3}$$

If $g(x)$ is equal to x, then the expectation value or the *true mean* of the random variable x is given by

$$m_x = E\left[x\right] = \int_a^b dx p(x) x, \qquad for \ \ a \leq x \leq b \tag{4.4}$$

The above formulation for a discrete random variable, x_i, of N outcomes reduces to

$$m_x = E[x_i] = \sum_{i=1}^{N} p(x_i)x_i, \qquad i = 1, N \tag{4.5}$$

Now, we define the expectation value of higher (k^{th}) powers of random variable x, referred to as the kth moment of x, by

$$m_x^k = E[x^k] = \int_a^b dx p(x)x^k \tag{4.6}$$

Similarly, for a discrete random variable x_i of N outcomes, the kth moment reduces to

$$m_x^k = E[x_i^k] = \sum_{i=1}^{N} p(x_i)x_i^k \tag{4.7}$$

Additionally, we define the kth central moment of continuous random variable x as

$$E[(x - m_x)^k] = \int_a^b dx p(x)(x - m_x)^k \tag{4.8}$$

Again, for a discrete random variable x_i of N outcomes, the central moment of kth order is expressed by

$$E[(x_i - m_x)^k] = \sum_{i=1}^{N} p(x_i)(x_i - m_x)^k \tag{4.9}$$

The central moment of order 2, i.e., $k = 2$, is referred to as the *true variance* of x, expressed by

$$\sigma_x^2 = E[(x - m_x)^2] = \int_a^b dx p(x)(x - m_x)^2 \tag{4.10}$$

Accordingly, for a discrete random variable, the variance formulation is given by

$$\sigma_x^2 = E[(x_i - m_x)^2] = \sum_{i=1}^{N} p(x_i)(x_i - m_x)^2 \tag{4.11}$$

The formulation for the variance can be written in a more convenient form if we expand the quadratic term as follows

$$\sigma_x^2 = \int_a^b dx p(x)(x^2 + m_x^2 - 2xm_x) =$$

$$\int_a^b dx p(x)x^2 + m_x^2 - 2m_x \int_a^b dx p(x)x;$$

$$\tag{4.12}$$

$$\sigma_x^2 = E\left[x^2\right] + m_x^2 - 2m_x^2;$$

$$\sigma_x^2 = E\left[x^2\right] - (E\left[x\right])^2.$$

Another useful quantity is the square root of the variance, referred to as the *true standard deviation*, i.e.

$$standard \; deviation \; of \; x \equiv \sigma_x = \sqrt{\sigma_x^2} \tag{4.13}$$

The *standard deviation* is an indication of the dispersion of random variable x relative to its mean (m_x).

The *true mean* and *true variance* also are referred to as the *population parameters*, because they are obtained based on a known probability density function, i.e., population.

4.2.2   Useful formulation for the expectation operator

Since the expectation operator, i.e., integration, is a linear operator, we can derive a number of useful identities which are beneficial when one is interested in evaluating:

- Expectation of a random variable or a function of a random variable is multiplied by a constant.

- Expectation of a linear combination of a few random variables or functions of random variables.

- Variance of a random variable or a function of a random variable is multiplied by a constant.

- Variance of a linear combination of a few random variables or functions of random variables.

Useful formulation to address the above situations are listed below:

$$E[ag(x) + b] = aE[g(x)] + b, \quad (4.14)$$

where a and b are constant coefficients.

$$E[ag(x_1) + bg(x_2)] = aE[g(x_1)] + bE[g(x_2)], \quad (4.15)$$

where x_1 and x_2 are two different random variables, and a and b are constant coefficients.

For a discrete random variable x_i of N outcomes

$$E\left[\sum_{i=1}^{N} a_i g(x_i)\right] = \sum_{i=1}^{N} a_i E[g(x_i)] \quad (4.16)$$

The variance formulation for a continuous random variable is given by

$$\sigma^2[ag(x)] = \int_a^b dx p(x)(ag(x) - a < g(x) >)^2 =$$
$$a^2 \int_a^b dx p(x)(g(x) - < g(x) >)^2; \quad (4.17)$$
$$\sigma^2[ag(x)] = a^2 \sigma^2[g(x)].$$

where $(<>)$ signs refer to the *mean* value of the function. Similarly, for a discrete random variable x_i of N outcomes, the variance is given by

$$\sigma^2\left[\sum_{i=1}^{N} a_i g(x_i)\right] = \sum_{i=1}^{N} \sigma^2[a_i g(x_i)] = \sum_{i=1}^{N} a_i^2 \sigma^2[g(x_i)]. \quad (4.18)$$

4.2.3 Multivariable

If a random variable is made of a combination of other random variables, e.g., a linear combination of two random variables, i.e.,

$$x_3 = c_1 x_1 + c_2 x_2 \quad (4.19)$$

Then, below we derive the variance of x_3 as follows:

$$\sigma^2(c_1 x_1 + c_2 x_2) \doteq E\left[(c_1 x_1 + c_2 x_2 - c_1 m_{x_1} - c_2 m_{x_2})^2\right]$$

$$\sigma^2(c_1x_1 + c_2x_2) = E\left[(c_1x_1 - c_1m_{x_1})^2\right] + E\left[(c_2x_2 - c_2m_{x_2})^2\right]$$
$$+ 2E\left[(c_1x_1 - c_1m_{x_1})(c_2x_2 - c_2m_{x_2})\right]$$

$$\sigma^2(c_1x_1 + c_2x_2) = c_1^2 E\left[(x_1 - m_{x_1})^2\right] + c_2^2 E\left[(x_2 - m_{x_2})^2\right]$$
$$+ 2c_1c_2 E\left[(x_1 - m_{x_1})(x_2 - m_{x_2})\right]$$

$$\sigma^2(c_1x_1 + c_2x_2) = c_1^2\sigma_{x_1}^2 + c_2^2\sigma_{x_2}^2 + 2c_1c_2 cov(x_1, x_2) \qquad (4.20)$$

where $cov(x_1, x_2) = E\left[(x_1 - m_{x_1})(x_2 - m_{x_2})\right]$.

Now, we can define the correlation coefficient between the two random variables as follows

$$\rho_{x_1,x_2} = \frac{cov(x_1, x_2)}{\sigma_{x_1}\sigma_{x_2}}$$

$$\rho_{x_1,x_2} = \frac{\int_{a_1}^{b_1} dx_1 \int_{a_2}^{b_2} dx_2 p(x_1, x_2)(x_1 - m_{x_1})(x_2 - m_{x_2})}{\sigma_{x_1}\sigma_{x_2}}$$

$$(4.21)$$

The two random variables, x_1 and x_2, may have two possible relations: independent; dependent.

If x_1 and x_2, are independent, then their combined *pdf*, $p(x_1, x_2)$, is given by

$$p(x_1, x_2) = p_1(x_1)p_2(x_2) \qquad (4.22)$$

Hence, the correlation coefficient formulation reduces to

$$\rho_{x_1,x_2} = \frac{\left(\int_{a_1}^{b_1} dx_1 p_1(x_1)(x_1 - m_{x_1})\right)\left(\int_{a_2}^{b_2} dx_2 p_2(x_2)(x_2 - m_{x_2})\right)}{\sigma_{x_1}\sigma_{x_2}}$$

$$\rho_{x_1,x_2} = \frac{\left(\int_{a_1}^{b_1} dx_1 p_1(x_1)x_1 - \int_{a_1}^{b_1} dx_1 p_1(x_1)m_{x_1}\right)}{\sigma_{x_1}} \times \qquad (4.23)$$
$$\frac{\left(\int_{a_2}^{b_2} dx_2 p_2(x_2)x_2 - \int_{a_2}^{b_2} dx_2 p_2(x_2)m_{x_2}\right)}{\sigma_{x_2}}$$

$$\rho_{x_1,x_2} = \frac{(m_{x_1} - m_{x_1})(m_{x_2} - m_{x_2})}{\sigma_{x_1}\sigma_{x_2}} = 0$$

This means that the formulation of the variance x_3 (Equation 4.19) reduces to

$$\sigma^2(c_1 x_1 + c_2 x_2) = c_1^2 \sigma_{x_1}^2 + c_2^2 \sigma_{x_2}^2 \tag{4.24}$$

Now, If x_1 and x_2, are dependent, e.g.,

$$x_1 = \alpha x_2 \tag{4.25}$$

Then, the correlation coefficient formulation reduces to

$$\rho_{x_1,x_2} = \frac{E\left[(\alpha x_2 - \alpha m_{x_2})(x_2 - m_{x_2})\right]}{\sigma_{x_1}\sigma_{x_2}} \tag{4.26}$$

$$\rho_{x_1,x_2} = \frac{\alpha \sigma_{x_2}^2}{\alpha \sigma_{x_2} \sigma_{x_2}} = 1$$

Therefore, if x_1 and x_2 are dependent on each other, then the variance of their linear combination, i.e., x_3 (Equation 4.19), reduces to

$$\sigma^2(c_1 x_1 + c_2 x_2) = c_1^2 \alpha^2 \sigma_{x_2}^2 + c_2^2 \sigma_{x_2}^2 + 2c_1 c_2 \alpha \sigma_{x_2}^2$$

$$\sigma^2(c_1 x_1 + c_2 x_2) = (c_1^2 \alpha^2 + c_2^2 + 2\alpha c_1 c_2)\sigma_{x_2}^2 \tag{4.27}$$

$$\sigma^2(c_1 x_1 + c_2 x_2) = (\alpha c_1 + c_2)^2 \sigma_{x_2}^2$$

4.3 SAMPLE EXPECTATION VALUES IN STATISTICS

The objective of any statistical method is to estimate averages and associated variances and confidence levels on the basis of a sampling process. In this section, we derive the formulations of the *sample mean* and the *sample variance*.

4.3.1 Sample mean

Based on the definition of the *true mean*, we define a formulation for the *sample mean* of sample size N as follows

$$\overline{x} = \frac{1}{N} \sum_{i=1}^{N} x_i \tag{4.28}$$

At the limit of an infinite number of samples, the above formulation has to approach the *true mean*. This means that we have to examine its *expectation* as follows:

$$E[\overline{x}] = E\left[\frac{1}{N}\sum_{i=1}^{N} x_i\right]$$

$$E[\overline{x}] = \frac{1}{N}\sum_{i=1}^{N} E\left[x_i\right] \tag{4.29}$$

$$E[\overline{x}] = \frac{1}{N}\sum_{i=1}^{N} m_x = \frac{1}{N}Nm_x = m_x$$

The above equality indicates that Equation 4.28 provides a good estimate of the *true mean*.

4.3.2 Sample variance

Based on the definition of variance given by Equation 4.11, we may define the *sample variance* as

$$s_x^2 = \frac{1}{N}\sum_{i=1}^{N}(x_i - \overline{x})^2 \tag{4.30}$$

Again, to verify the validity of the above formulation, we derive its expectation value as follows

$$E\left[s_x^2\right] = E\left[\frac{1}{N}\sum_{i=1}^{N}(x_i - \overline{x})^2\right] \tag{4.31}$$

Now, if add and subtract m_x in the the right-hand side of the above equation, we obtain

$$E\left[s_x^2\right] = E\left[\frac{1}{N}\sum_{i=1}^{N}(x_i - m_x + m_x - \overline{x})^2\right]$$

$$= \frac{1}{N}\sum_{i=1}^{N} E\left[(x_i - m_x)^2\right] + \frac{1}{N}\sum_{i=1}^{N} E\left[(\overline{x} - m_x)^2\right]$$

$$+ 2E\left[(m_x - \overline{x})\frac{1}{N}\sum_{i=1}^{N}(x_i - m_x)\right]$$

or

$$E\left[s_x^2\right] = \frac{1}{N} \sum_{i=1}^{N} \sigma_x^2 + \frac{1}{N} N \times E\left[(\overline{x} - m_x)^2\right] - 2E\left[(\overline{x} - m_x)^2\right]$$

or

$$E\left[s_x^2\right] = \sigma_x^2 - E\left[(\overline{x} - m_x)^2\right] \tag{4.32}$$

The second term on the right-hand side of the above equation is the variance of the sample average $\overline{x}$. Hence, we can obtain its formulation as follows

$$\sigma_{\overline{x}}^2 = \sigma^2(\overline{x}) = \sigma^2 \left(\frac{1}{N} \sum_{i=1}^{N} x_i\right) = \sum_{i=1}^{N} \frac{1}{N^2} \sigma^2(x_i)$$

or

$$\sigma_{\overline{x}}^2 = \frac{1}{N^2} N \sigma_x^2 = \frac{\sigma_x^2}{N} \tag{4.33}$$

The above formulation provides an important information that the variance of the average reduces with increasing number of samples. This will be addressed further later in this chapter.

Now, if we substitute Equation 4.33 into Equation 4.32, we obtain

$$E\left[s_x^2\right] = \frac{N-1}{N} \sigma_x^2 \tag{4.34}$$

The above equation demonstrates that the sample variance formulation (Equation 4.28) is a biased estimate of the *true variance*. Hence, to define an unbiased formulation for the sample variance, we rewrite above equation as

$$E\left[\frac{N}{N-1} s_x^2\right] = \sigma_x^2 \tag{4.35}$$

This means that indeed the term in the bracket on the left-hand side of the above equation yields an unbiased estimate of the *true variance*. Hence, the unbiased sample variance (S_x^2) is given by

$$S_x^2 = \frac{N}{N-1} s_x^2 \tag{4.36}$$

Then, if we substitute Equation 4.28 into the above equation, the *unbiased sample variance* formulation reduces to

$$S_x^2 = \frac{N}{N-1} \times \frac{1}{N} \sum_{i=1}^{N} (x_i - \overline{x})^2 = \frac{1}{N-1} \sum_{i=1}^{N} (x_i - \overline{x})^2 \tag{4.37}$$

4.4   PRECISION AND ACCURACY OF A SAMPLE AVERAGE

To determine the confidence level on the outcome of a statistical process, it is necessary to determine the precision and accuracy associated with the sample average. The precision is measured by the *relative statistical uncertainty* that is defined by

$$R_x = \frac{\sigma_x}{\overline{x}} \tag{4.38}$$

Note that the *standard deviation* (σ_x) is also referred to as the *statistical uncertainty*.

The accuracy refers to the measure of the deviation from the *truth*, i.e., how far is the precise average from the *true mean*? Generally, to estimate the accuracy, it is necessary to either perform an experiment or to compare with the results of another formulation/technique that is known to be accurate.

Thus far, we have introduced the formulations of the *true mean* and *variance* and *sample mean* and *variance*. In the remainder of this chapter, we will introduce techniques for estimating the confidence level on the estimated sample average. To do this, we will introduce a few density functions commonly used for representation of most random physical processes, and their associated limit theorems for estimating the *confidence level*.

4.5   COMMONLY USED DENSITY FUNCTIONS

In this section, we will introduce a few density functions commonly encountered when dealing with various random physical processes.

4.5.1   Uniform density function

A uniform density function $f(x)$ is constant within the range of a random variable x. For example, if x is defined in a range of $[a, b]$, then $f(x)$ is given by

$$f(x) = k \tag{4.39}$$

Then, we derive the corresponding *pdf* as follows

$$p(x) = \frac{f(x)}{\int_a^b dx f(x)} = \frac{k}{k(b-a)} = \frac{1}{b-a} \tag{4.40}$$

Then, the *true mean* of the random variable x is given by

$$m_x = \int_a^b dx\, x p(x) = \int_a^b dx \frac{1}{b-a} x = \frac{a+b}{2} \tag{4.41}$$

Similarly, the *true variance* is derived as follows

$$\sigma_x^2 = E[x^2] - m_x^2 = \int_a^b dx\, p(x) x^2 - \left(\frac{a+b}{2}\right)^2 \tag{4.42}$$

$$\sigma_x^2 = \int_a^b dx \frac{1}{b-a} x^2 - \left(\frac{a+b}{2}\right)^2 = \frac{(b-a)^2}{12}$$

4.5.2 Binomial density function

In order to introduce the binomial density function, it is necessary to introduce the Bernoulli random process and its associated *pdf*.

4.5.2.1 Bernoulli process

A Bernoulli process refers to a random process that has only two outcomes and the probabilities of these outcomes remain constant throughout the experimentation. The probability density function of a Bernoulli process is given by

$$p(n) = p^n (1-p)^{1-n}, \qquad for \ n = 0, \ 1 \tag{4.43}$$

where n refers to the outcome (or random variable), and p refers to the probability of one of the outcomes, e.g., *success*. There are many examples for a Bernoulli process; e.g., a coin toss is a Bernoulli process.

The *true mean* for a Bernoulli random variable is given by

$$m_n = E[n] = \sum_{n=0}^{1} n p(n) = \sum_{n=0}^{1} n p^n (1-p)^{1-n} = 0 + p = p \tag{4.44}$$

And, the *true variance* for a Bernoulli random variable is given by

$$\sigma_n^2 = E[n^2] - m_n^2 = \sum_{n=0}^{1} n^2 p^n (1-p)^{1-n} - p^2 = p - p^2 = p(1-p) = pq \tag{4.45}$$

where $q = 1 - p$ and refers to the probability of *failure*.

4.5.2.2 *Derivation of the Binomial density function*

Consider a Bernoulli random process is repeated N times, with outcomes (n_i's), then the sum of these outcomes ($n = \sum_i^N n_i$) follows a binomial distribution. Here, we derive the formulation of the binomial distribution; given the probability of *success* (or outcome of interest) is p, then the probability of n successes out of N experiments is given by

$$p(n) = C_N(n)p^n q^{N-n} \qquad (4.46)$$

where $C_N(n)$ refers to the number of combinations of N Bernoulli experiments ($n_i = 0$ or 1) that their sum is equal to $n = \sum_{i=1}^N n_i$, irrespective of the order of successful experiment. The number of combinations of experiments with a sum n is determined as follows:

- The number of permutations of N outcomes that their sum is equal to n is equal to:

$$N(N-1)(N-1)\cdots(N-n+1) = \frac{N!}{(N-n)!}; \qquad (4.47)$$

- Since achieving a sum of n is not dependent on the order of successful outcomes in N experiments; therefore, the number of unique combinations is obtained by removing the number of permutations of n successful outcomes, i.e.,

$$C_N(n) = \frac{N!}{(N-n)!} \times \frac{1}{n!} = \frac{N!}{(N-n)!n!} \qquad (4.48)$$

Therefore, a binomial density function is expressed by

$$p(n) = \frac{N!}{(N-n)!n!}p^n q^{N-n}, \qquad for \ \ n = 1, N \qquad (4.49)$$

The above density function determines the probability of obtaining n successful outcomes out of N experiments (trials).

The expectation value of n, or the *true mean* is given by

$$m_n = E[n] = E\left[\sum_{i=1}^N n_i\right] = \sum_{i=1}^N E[n_i] = Np \qquad (4.50)$$

And the variance of n, or the *true variacne* is given by

$$\sigma_n^2 = \sigma^2(n) = \sigma^2\left[\sum_{i=1}^N n_i\right] = \sum_{i=1}^N \sigma^2[n_i] = Npq \qquad (4.51)$$

Hence, the relative uncertainty is given by

$$R_n = \frac{\sigma_n}{m_n} = \frac{\sqrt{Npq}}{Np} = \sqrt{\frac{q}{Np}} \tag{4.52}$$

This above formulation indicates that the relative uncertainty decreases with increasing number of experiments.

To determine $p(n)$ or $P(n)$ for a binomial distribution, we can utilize the recursive formulations given in Table 4.1

Now, using the algorithm given in Table 4.2, we determine the *pdf* and *cdf* for a binomial distribution with $p = 0.7$. Figures 4.1 and 4.2 show the distributions for N equal to 20 and 100, respectively. It is worth noting that, as expected, the maximum value of the *pdf*s occurs at the mean values ($m = Np$), i.e., 14 and 70, corresponding to N equal to 20 and 100, respectively.

Table 4.1 Recursive formulations for determination of the binomial distribution and its cumulative density function

pdf	cdf
$p(0)=(1-p)^n$	$P(0)=p(0)$
$p(n)=\frac{p}{1-p} \times \frac{N-n+1}{n}p(n-1)$	$P(n)=\sum_{n'=0}^{n} p(n')$

Table 4.2 An algorithm for determination of the binomial distribution *pdf* and *cdf*

Algorithm	Description
$f(1) = (1-p)^n$	probability of 0 success
DO i=1,N+1	
$\quad f(i)=\frac{p}{1-p} \times \frac{n-i+2}{i-1}f(i-1)$	probability of 1 to N successes
ENDDO	
tf(1)=f(1)	CDF for 0 success
DO i=2,N+1	
$\quad tf(i)=tf(i-1)+f(i)$	CDF for success from 1 to N
ENDDO	

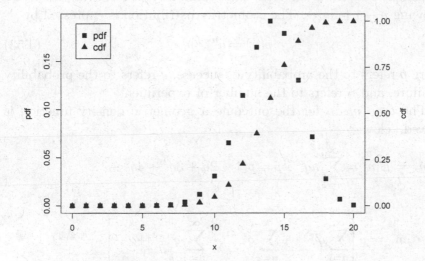

Figure 4.1 Schematic of a binomial *pdf* and *cdf* with $p = 0.7$ and $N = 20$

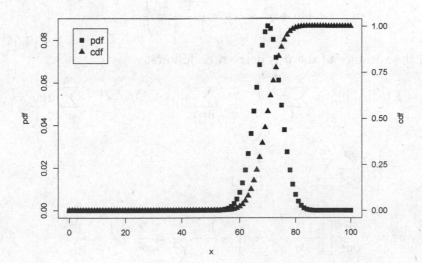

Figure 4.2 Schematic of a binomial *pdf* and *cdf* with $p = 0.7$ and $N = 100$

4.5.3 Geometric density function

A geometric distribution gives the probability of achieving a success following $n-1$ failures. The geometric distribution is expressed by

$$p(n) = q^{n-1}p, \tag{4.53}$$

where p refers to the probability of success, q refers to the probability of failure, and n refers to the number of experiments.

The *true mean* for the outcome a geometric density function is derived below:

$$m_n = E[n] = \sum_{n=1}^{\infty} nq^{n-1}p = p(1 + 2q + 3q^2 + 4q^3 + \dots)$$

$$m_n = p \left[\sum_{n=1}^{\infty} q^{n-1} + \sum_{n=2}^{\infty} q^{n-1} + \sum_{n=3}^{\infty} q^{n-1} + \cdots \right]$$

$$m_n = p \left[\frac{1}{1-q} + \frac{q}{1-q} + \frac{q^2}{1-q} + \cdots \right]$$

$$m_n = 1 + q + q^2 + \cdots = \frac{1}{1-q} = \frac{1}{p}$$

$$\tag{4.54}$$

And the variance of the n is derived as follows:

$$\sigma_n^2 = E[n^2] - m_n^2 = \sum_{n=1}^{\infty} n^2 pq^{n-1} = \sum_{n=1}^{\infty} n(n+1)pq^{n-1} - \sum_{n=1}^{\infty} npq^{n-1}$$

$$\sigma_n^2 = p \frac{\partial^2}{\partial q^2} \sum_{n=1}^{\infty} q^{n+1} - \frac{1}{p} - \frac{1}{p^2}.$$

$$\sigma_n^2 = p \frac{\partial^2}{\partial q^2} \left[\frac{1}{1-q} - 1 - q \right] - \frac{1}{p} - \frac{1}{p^2} = \frac{2}{p^2} - \frac{1}{p} - \frac{1}{p^2}$$

$$\sigma_n^2 = \frac{1}{p^2} - \frac{1}{p}$$

$$\tag{4.55}$$

This distribution can be used to estimate the average number of collisions a particle may go through in a homogeneous infinite medium before a specific interaction occurs, given that the relative probability of different outcomes is constant.

4.5.4 Poisson density function

The binomial distribution approaches a Poisson distribution if we have the following conditions:

$$p \ll 1 \tag{4.56}$$

$$N \gg 1 \tag{4.57}$$

and

$$n \ll N \tag{4.58}$$

Now, we may apply the above conditions to a binomial distribution to obtain the Poisson distribution. We may rewrite the binomial distribution, Equation 4.49 as

$$p(n) = \frac{N \times (N-1) \times (N-2) \cdots (N-n+1)}{n!} p^n (1-p)^{N-n} \tag{4.59}$$

If we apply Equation 4.58, for simplicity replace m_n with m, and substitute p with $\frac{m}{N}$, the above equation reduces to

$$p(n) = \frac{N^n}{n!} \left(\frac{m}{N}\right)^n \left(1 - \frac{m}{N}\right)^{N-n} \tag{4.60}$$

$$p(n) = \frac{N^n}{n!} \left(1 - \frac{m}{N}\right)^{-n} \left(1 - \frac{m}{N}\right)^{N}$$

Again, using Equation 4.58, the second term on the right-hand side of the above equation reduces to

$$\left(1 - \frac{m}{N}\right)^{-n} = 1 + n\left(\frac{m}{N}\right) + \frac{-n(-n-1)}{2!}\left(\frac{m}{N}\right)^2 + \cdots \tag{4.61}$$

$$\left(1 - \frac{m}{N}\right)^{-n} = 1 + m\left(\frac{n}{N}\right) + \frac{m^2}{2!}\left(\frac{n(n+1)}{N^2}\right)^2 + \cdots = 1,$$

and because of Equation 4.57, the third term on the right-hand side of Equation 4.60 reduces to

$$lim_{N \to \infty} \left(1 - \frac{m}{N}\right)^N = e^{-m} \tag{4.62}$$

Hence, $p(n)$ reduces to

$$p(n) = \frac{m^n}{n!} e^{-m} \qquad (4.63)$$

The above equation is referred to as the *Poisson* density function.

Now, we derive the formulation for the *true mean* of n as follows

$$E[n] = \sum_{n=0}^{N} np(n) = \sum_{n=0}^{N} n \frac{m^n}{n!} e^{-m} = me^{-m} \sum_{n=1}^{N} \frac{m^{n-1}}{(n-1)!}, \qquad (4.64)$$

Note that the minimum value of the above summation is set to 1, as the term is zero for $n = 0$. After a change of variable, i.e., $k = n - 1$, we obtain

$$E[n] = me^{-m} \sum_{k=0}^{N-1} \frac{m^k}{k!} \qquad (4.65)$$

Now, if we consider the fact that the higher terms of the above series become negligible, we may set its upper limit to infinity, then the series is simply an exponential function, hence, the expected value reduces to

$$E[n] = me^{-m} \sum_{k=0}^{\infty} \frac{m^k}{k!} = me^{-m} e^m = m \qquad (4.66)$$

Now, we derive the formulation for the variance of n as follows

$$\sigma_n^2 = E[n^2] - m^2 = \sum_{n=0}^{\infty} \frac{m^n}{n!} e^{-m} n^2 - m^2 = e^{-m} \sum_{n=0}^{\infty} \frac{m^n}{n!} n^2 - m^2$$

$$\sigma_n^2 = me^{-m} \sum_{n=0}^{\infty} \frac{m^{n-1}}{(n-1)!} n - m^2 = m$$

$$(4.67)$$

Hence, the relative uncertainty is given by

$$R_n = \frac{\sqrt{m}}{m} = \frac{1}{\sqrt{m}} \qquad (4.68)$$

To determine $p(n)$ or $P(n)$ for a *Poisson* density function, we can utilize the recursive formulations given in Table 4.3.

Table 4.3 Recursive formulations for determination of the Poisson distribution and its cumulative density function

pdf	cdf
$p(0)=e^{-m}$	$P(0)=p(0)$
$p(n)=\frac{m}{n}p(n-1)$	$P(n)=\sum_{n'=0}^{n} p(n')$

4.5.5   Normal (*Gaussian*) density function

For a large number of experiments (N), the de Moivre–Laplace limit theorem states that a *binomial* density function approaches a normal (Gaussian) density function that is a continuous function.

To derive the normal distribution, we find the logarithm of the binomial density function, Equation 4.49 as

$$\ln p(n) = \ln(N!) - \ln(n!) - \ln\left[(N-n)!\right] + n\ln(p) + (N-n)\ln(q) \quad (4.69)$$

Now, we will find the value of n for which $\ln[p(n)]$ is maximum, i.e.,

$$\frac{d(p(n))}{dn} = 0 \tag{4.70}$$

Then, considering the following identity

$$\frac{d(\ln(n!))}{dn} \simeq \ln(n), \qquad for \ n \gg 1, \tag{4.71}$$

Equation 4.69 reduces to

$$-\ln(n) + \ln(N-n) - \ln(p) - \ln(q) = 0$$

$$\tag{4.72}$$

$$\ln\left[\frac{(N-n)p}{nq}\right] = 0$$

This means that

$$\frac{(N-n)p}{nq} = 1 \tag{4.73}$$

Hence, this means that

$$(N - n)p = nq$$

$$(4.74)$$

$$Np = n(p + q)$$

Since $p + q$ is equal 1, then the value of n is given by

$$\tilde{n} = Np \qquad (4.75)$$

Now, we investigate the behavior of $\ln[p(n)]$ near its maximum at $\tilde{n}$. For this, we expand $\ln[p(n)]$ in a Taylor series about $\tilde{n}$. This means that

$$\ln[p(n)] = \ln[p(\tilde{n})] + \left[\frac{d(\ln(p(n)))}{dn}\right]_{\tilde{n}} (n - \tilde{n}) +$$

$$\frac{1}{2!}\left[\frac{d^2(\ln(p(n)))}{dn^2}\right]_{\tilde{n}} (n - \tilde{n})^2 + \qquad (4.76)$$

$$\frac{1}{3!}\left[\frac{d^3(\ln(p(n)))}{dn^3}\right]_{\tilde{n}} (n - \tilde{n})^3 + \cdots$$

In the above equation, the first derivative at $\tilde{n}$ is equal zero and terms that involve $(n - \tilde{n})^3$ and higher powers are negligible. This means that the above equation reduces to

$$\ln[p(n)] = \ln[p(\tilde{n})] + \frac{1}{2!}\left[\frac{d^2(\ln(p(n)))}{dn^2}\right]_{\tilde{n}} (n - \tilde{n})^2 \qquad (4.77)$$

Now, we determine the 2nd order derivative term by finding the derivative of Equation 4.69 as

$$\frac{d^2(\ln(p(n)))}{dn^2} = \frac{1}{n} + \frac{1}{N - n} = -\frac{N}{n(N - n)}. \qquad (4.78)$$

For $n = \tilde{n}$, the above equation reduces to

$$\left[\frac{d^2(\ln(p(n)))}{dn^2}\right]_{\tilde{n}} = \frac{N}{\tilde{n}(N - \tilde{n})} = -\frac{N}{Np(N - Np)} = -\frac{1}{Npq} \qquad (4.79)$$

Then, Equation 4.77 reduces to

$$\ln\left[\frac{p(n)}{p(\tilde{n})}\right] = -\frac{(n - \tilde{n})^2}{2Npq}$$

$$(4.80)$$

$$p(n) = p(\tilde{n})e^{-\frac{(n - \tilde{n})^2}{2Npq}}$$

To evaluate the constant coefficient, $p(\tilde{n})$, we require that

$$\sum_{n=1}^{N} p(n) = 1 \tag{4.81}$$

Since $p(n)$ changes only slightly between successive integral values of n, this sum can be replaced by an integration as follows

$$\int_{-\infty}^{\infty} dn p(\tilde{n}) e^{-\frac{(n-\tilde{n})^2}{2Npq}} = 1 \tag{4.82}$$

Now, considering the following identity

$$\int_{-\infty}^{\infty} dx e^{-\alpha x^2} = \sqrt{\pi} \alpha^{-\frac{1}{2}} \tag{4.83}$$

Using above identity in Equation 4.82, we obtain

$$p()\sqrt{\pi}\sqrt{2Npq} = 1 \tag{4.84}$$

$$p(\tilde{n}) = \frac{1}{\sqrt{2\pi Npq}}$$

Hence, the formulation for the *normal* density function is given by

$$p(n) = \frac{1}{\sqrt{2\pi Npq}} e^{-\frac{(n-Np)^2}{2Npq}} \tag{4.85}$$

For a binomial density function, the *true mean* and *variance* are given by $m = Np$ and $\sigma^2 = Npq$, respectively. Using these formulations in the above equation, we obtain the formulation of the *normal* density function for random variable x as follows

$$p(x) = \frac{1}{\sqrt{2\pi\sigma^2}} e^{-\frac{(x-m)^2}{2\sigma^2}} \tag{4.86}$$

To examine the accuracy of a normal density function for approximating a binomial density function, we consider a Bernoulli process with a probability of success (p) equal to 0.1. Figure 4.3 compares probabilities predicted by binomial and normal density functions for five trials. As expected, for so few trials, the outcomes of the two distributions are very different and this difference becomes smaller as we seek a larger number of successes.

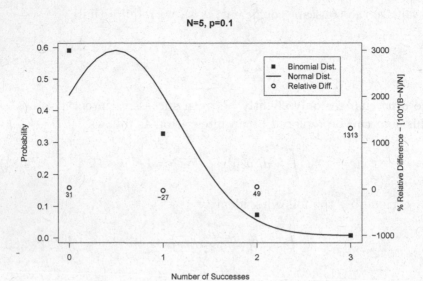

Figure 4.3   Comparison of normal and binomial distribution for five trials

If we increase the number of trials, we observe closer agreements between the predictions of the two distributions. For example, Figure 4.4 shows a case with 100 trials that results in a significantly closer agreement, differences <11% in the range of 4 to 16.

Figures 4.3 and 4.4 demonstrated that as expected a normal distribution becomes more suitable for representation of a binomial distribution with larger number of experiments.

To discuss the characteristics of a normal density function, in Figure 4.5 we show a normal distribution with its *cdf* with a *mean(m)* of 40 and a *standard deviation(σ)* of 10.

Major characteristics of a normal density function (e.g., Figure 4.5) are listed below:

- Peaks at the mean.

- It is symmetric about the mean.

- Points of maximum slope, i.e., $\frac{d^2(p(x))}{dx^2} = 0$, points of inflection occur at $\pm\sigma$ relative to the mean. At these points, the value of the distribution is ~ 60% of its maximum value. (Note that the slope is equal to $\frac{0.242}{2\sigma^2}$).

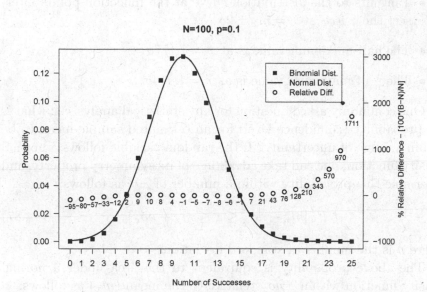

Figure 4.4 Comparison of normal and binomial distribution for 100 trials

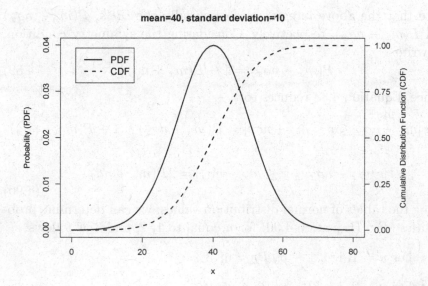

Figure 4.5 Schematic of a normal density function and its *cdf*

- Tangents to the distribution curve at the inflection points intersect the *x-axis* at $x = m_x \pm 2\sigma$.

- The half-maximum value is at $x = 1.177\sigma$.

- The $\frac{1}{e}$ of maximum value is at $x = 1.414\sigma$.

One commonly asked question for any statistical analysis is: what is the probability (confidence level) for an estimated sample mean to be within a range of uncertainty? If the random variable follows a normal density function, one can take advantage of its symmetry property, and determine this probability within a number of $\pm\sigma$ as follows:

$$Pr\left[m_x - n\sigma_x \leq x \leq m_x + n\sigma_x\right], \tag{4.87}$$

where n is the number of standard deviations.

The above probability is equivalent to the area under a normal density function within $\pm n\sigma_x$ relative to the mean (m_x) as follows:

$$Pr\left[m_x - n\sigma_x \leq x \leq m_x + n\sigma_x\right] =$$
$$\frac{1}{\sqrt{2\pi\sigma_x^2}}\left[\int_{-\infty}^{m_x+n\sigma_x} dx e^{-\frac{(x-m_x)^2}{2\sigma_x^2}} - \int_{-\infty}^{m_x-n\sigma_x} dx e^{-\frac{(x-m_x)^2}{2\sigma_x^2}}\right] \tag{4.88}$$

Note that the above integrals are essentially the *cdf's*, $P(m_x + n\sigma_x)$ and $P(m_x - n\sigma_x)$, respectively. Considering the symmetry condition, we write

$$P(m_x - n\sigma_x) = 1 - P(m_x + n\sigma_x) \tag{4.89}$$

Hence, Equation 4.88 reduces to

$$Pr\left[m_x - n\sigma_x \leq x \leq m_x + n\sigma_x\right] = P(m_x + n\sigma_x) - 1 + P(m_x + n\sigma_x)$$

$$Pr\left[m_x - n\sigma_x \leq x \leq m_x + n\sigma_x\right] = 2P(m_x + n\sigma_x) - 1 \tag{4.90}$$

Using the tables of normal distribution values, we can determine probabilities (Pr) (Equation 4.90) for n equal to 1, 2, and 3 as follows:

- For $n = 1$, i.e., $1 - \sigma$, $Pr = 68.3\%$

- For $n = 2$, i.e., $2 - \sigma$, $Pr = 95.4\%$

- For $n = 2$, i.e., $3 - \sigma$, $Pr = 99.7\%$

The above probability values indicate that $\sim 68.3\%$, $\sim 95.4\%$, and $\sim 99.7\%$ of random variables lie within *one*, *two*, and *three* standard deviations from the *mean*, respectively. Note that these percentages are true for any normal distribution because the distribution is normalized.

A simpler form of normal distribution can be derived if we consider a change of variable as

$$t = \frac{x - m_x}{\sigma_x} \tag{4.91}$$

Then, since t and x have a one-to-one relation, we may derive a density function in terms of variable t as follows

$$|\phi(t)dt| = |p(x)dx| \tag{4.92}$$

Considering that the two density functions are positive quantities, then we can solve for $\phi(t)$ as follows

$$\phi(t) = p(x)\left|\frac{dx}{dt}\right| = p(x)\sigma \tag{4.93}$$

$$\phi(t) = \frac{1}{\sqrt{2\pi}}e^{-\frac{t^2}{2}}$$

Now, we determine the *true mean* and *true variance* for the new random variable t as follows:

true mean for the t random variable:

$$m_t = E[t] = E\left[\frac{x - m_x}{\sigma_x}\right] = \frac{E[x] - m_x}{\sigma_x} = \frac{m_x - m_x}{\sigma_x} = 0 \tag{4.94}$$

true variance for the t random variable:

$$\sigma_t^2 = \sigma^2\left[\frac{x - m_x}{\sigma_x}\right] = \frac{1}{\sigma_x^2}\sigma_x^2 = 1 \tag{4.95}$$

4.6   LIMIT THEOREMS AND THEIR APPLICATIONS

In this section, we will discuss two limit theorems that are commonly used for estimating the confidence level on the results of a statistical analysis.

4.6.1 Corollary to the de Moivre-Laplace limit theorem

To derive this corollary, we want to determine the probability (Pr) that the *sampled* probability of success of a Bernoulli process (p') is within a certain range (ϵ) of the *true* probability of success (p) as follows

$$Pr\left[|p' - p| \le \epsilon\right] = ? \tag{4.96}$$

To respond to the above question mark, we expand the above inequality as follows

$$\left|\frac{x}{N} - p\right| \le \epsilon, \tag{4.97}$$

where x refers to the number of successes in N experiments. Now, we divide both sides of the above inequality by $\sqrt{Npq}$, and rearrange terms to obtain

$$\left|\frac{x - Np}{\sqrt{Npq}}\right| \le \epsilon\sqrt{\frac{N}{pq}}, \tag{4.98}$$

Note that the de Moivre-Laplace theorem states that at the limit of large number of experiments (N), the binomial distribution is represented by a normal distribution for the x random variable with the true mean of $m_x = Np$, and the true variance of $\sigma_x^2 = Npq$. Hence, Equation 4.98 can be written as

$$\left|\frac{x - m_x}{\sigma_x}\right| \le \epsilon\sqrt{\frac{N}{pq}} \tag{4.99}$$

Since the random variable t is equal to $\frac{x - m_x}{\sigma_x}$, then above inequality can be written as

$$|t| \le \epsilon\sqrt{\frac{N}{pq}}, \tag{4.100}$$

Then, the probability that the above inequality is true can be obtained by using the normal density function for the random variable t as follows

$$Pr\left[|t| \le \epsilon\sqrt{\frac{N}{pq}}\right] = 2\Phi\left(\epsilon\sqrt{\frac{N}{pq}}\right) - 1 \tag{4.101}$$

Hence, the *corollary to the de Moivre-Laplace limit theorem*, i.e., Equation 4.96, is expressed by

$$Pr\left[|p' - p| \le \epsilon\right] = 2\Phi\left(\epsilon\sqrt{\frac{N}{pq}}\right) - 1 \tag{4.102}$$

To demonstrate the usefulness of this corollary, let's consider the following example:

Example 1 - For a Bernoulli random variable, after $N = 100$ experiments, $x = 8$ successes are achieved. What is the level of confidence that the sampled probability is within $\epsilon = \pm 1.0\%$ of the true probability? Using Equation 4.102 for $p' = \frac{8}{100}$, we can estimate the Pr as follows

$$Pr\left[|.08 - p| \leq .01\right] = 2\Phi\left[.01\sqrt{\frac{100}{.08 \times 0.92}}\right] - 1 = \tag{4.103}$$
$$2 \times 0.644 - 1 = 0.29$$

This means the corollary indicates that the confidence level for this experiment is only 29%!

Example 2 - How many experiments (N) are needed in order to achieve a 95% confidence on an ϵ of 0.2% for a Bernoulli random variable? This means that

$$Pr\left[|p' - p| \leq .002\right] = 0.95$$

$$2\Phi\left[\epsilon\sqrt{\frac{N}{pq}}\right] - 1 = .95 \tag{4.104}$$

$$\Phi\left[\epsilon\sqrt{\frac{N}{pq}}\right] = 0.975$$

Using tables of the normal density function,

$$\epsilon\sqrt{\frac{N}{pq}} = 1.96 \tag{4.105}$$

Hence, N is given by

$$N = \left(\frac{1.96}{\epsilon}\right)^2 pq \simeq 10^6 pq \tag{4.106}$$

To be conservative, we consider the maximum value of pq that is $\frac{1}{4}$ corresponding to $p = q = \frac{1}{2}$, hence

$$N = 250,000 \tag{4.107}$$

Assuming a maximum value for pq results in a large number of experiments (250,000) that may require a significant amount of computer time. However, it is recommended to conduct initial sampling to estimate p', and therefore avoid unnecessary waste of resources. For example, if the estimated p' is equal to 0.01, then the corollary predicts a reduced number of experiments of only about 10,000 that is smaller by a factor of 25! This is true for the corollary that has been derived based on the absolute difference of the *sampled* and *true* probabilities of success.

To examine the precision of a sample average, however, it is necessary to examine its relative uncertainty defined as $\frac{S_{\bar{x}}}{\bar{x}}$. In the second example, the two cases considered have relative uncertainties (or precision) of 0.4% and 20%, respectively. This means we have the same 95% confidence for two aforementioned precisions that differ by a factor 50!

To avoid the observation made in the second example, we may derive the formulation for the corollary based on the relative difference of the sample and true *means* as follows:

$$\left| \frac{p' - p}{p} \le \epsilon \right| \tag{4.108}$$

Then, if we divide both sides of the above inequality by $\sqrt{Npq}$, and rearrange the terms, we obtain the following

$$\left| \frac{x - m_x}{\sigma_x} \right| \le \epsilon \sqrt{\frac{Np}{q}} \tag{4.109}$$

Hence, the probability for the relative difference of the sampled and true success probabilities is expressed by

$$Pr \left[\left| \frac{p' - p}{p} \right| \le \right] = 2\Phi \left(\sqrt{\frac{Np}{q}} \right) - 1 \tag{4.110}$$

Example 3 - Now, if we use Equation 4.110, the predicted number of experiments is given by

$$N \simeq 10^6 \left(\frac{q}{p} \right) \tag{4.111}$$

Hence, for the two estimated probabilities of success, the predicted number of experiments are:

$p' = \frac{1}{2}$, $N \simeq 10^6 \left(\frac{.5}{.5}\right) = 10^6$, and

$p' = 0.01$, $N \simeq 10^6 \left(\frac{0.99}{0.01}\right) = 10^8$

The above predicted number of experiments is more realistic, because intuitively more experiments are needed when the probability of success is smaller.

4.6.2 Central limite theorem

Consider x_1, x_2, $\cdots$, x_N are the outcomes of N independent samples from a common density function and the mean (m_x) and the variance (σ_x^2) of the distribution exist; then, for any fixed value of N histories per trial, there is a *pdf* $(f_N(x))$, which describes the distribution of x that results from the repeated trials. As N approaches infinity, the *central limit theorem* states that there is a limiting density function for x that is a *normal density function* given by

$$f_N(\overline{x}) = \frac{1}{\sqrt{2\pi\sigma_{\overline{x}}^2}} e^{-\frac{(\overline{x}-m_x)}{2\sigma_{\overline{x}}^2}}, \qquad as\ N \to \infty \qquad (4.112)$$

where $\sigma_{\overline{x}}^2 = \frac{\sigma_x^2}{N}$ (Equation 4.33).

Provided that N is sufficiently large, we may use the above formulation to estimate a reduced variance for $\overline{x}$. Commonly, we cannot estimate σ_x^2 exactly, hence, we approximate it by calculating the sample variance (S_x^2).

Again, we may estimate the confidence level on the estimated uncertainty $(\sigma_{\overline{x}})$ as follows:

1. Considering a change of variable of $\overline{t} = \frac{\overline{x}-m_x}{\sigma_{\overline{x}}}$, then

2. Probability that $\overline{t}$ is within n standard deviations $(\sigma_{\overline{x}})$ is given by

$$Pr\left[-n \leq \overline{t} \leq n\right] = 2\Phi(n) - 1 \qquad (4.113)$$

Again, the confidence levels for n=1, 2, and 3 are 68.3%, 95.4%, and 99.7%, respectively.

The significance of the central limit theorem is the fact that it is applicable to any random variable which has a well-defined m_x and σ_x^2. Moreover, it emphasizes on the uncertainty of the *sample mean* $(\overline{x})$ rather than the random variable (x) itself.

4.6.2.1 Demonstration of the Central Limit Theorem

Given we perform a Monte Carlo simulation consisting of T trials of N experiments/trial, what are the sample mean and the sample variance?

The sample mean is obtained via the following two steps:

1. The sample mean per trial k is given by

$$\overline{x}_k = \frac{1}{N} \sum_{i=1}^{N} x_{ki}. \tag{4.114}$$

2. The sample mean for all trials is given by

$$< \overline{x}_k > = \frac{1}{T} \sum_{k=1}^{T} \overline{x}_k, \qquad k = 1, T \tag{4.115}$$

The sample variance is obtained by

$$S_x^2 = \frac{1}{T-1} \sum_{k=1}^{T} (\overline{x}_k - < \overline{x}_k >)^2 \tag{4.116}$$

Now if we assume that $\overline{x}_k$ is normally distributed about ¡$\overline{x}_k$¿, then we may state that the *true mean* (m_x) is within

$$| < \overline{x}_k > -m_x | \le n \cdot S_x \tag{4.117}$$

with confidence levels of 68.3%, 95.4%, and 99.7% for n equal to 1, 2, and 3, respectively.

If the random variable $\overline{x}_k$ is normally distributed about the sample mean, then it is expected that $< \overline{x} >$ is normally distributed about the true mean with a variance of $\frac{S_x^2}{T}$ according to the central limit theorem. Even though we have calculated only one value of $< \overline{x} >$, we may state that the *true mean* (m_x) is within

$$| < \overline{x}_k > -m_x | \le n \cdot \frac{S_x}{\sqrt{T}} \tag{4.118}$$

with confidence levels of 68.3%, 95.4%, and 99.7% for n equal to 1, 2, and 3, respectively.

The above discussion can be demonstrated by using the following simple example.

Table 4.4 Demonstration of the Central Limit Theorem

Experiments Trial(N)	Trials (T)	Sample mean $(\overline{x})$	sample variance (σ_x^2)
1	1000	0.498	0.249
2	500	0.498	0.119
4	250	0.498	0.060
8	125	0.498	0.0307
10	100	0.498	0.0252
40	25	0.498	0.00958
100	10	0.498	0.00344
200	5	0.498	0.00078
500	2	0.498	0.000061

Example - Let's consider a random variable x, with $m_x = 0.5$ and $\sigma_x^2 = 0.25$. If the corresponding FFMC is given by

$$x = -\frac{\ln \eta}{2}, \tag{4.119}$$

then using a computer code we perform 1000 samples of x. Then we estimate the sample average and sample variance for different grouping of the 1000 experiments into *Experiments per Trial* and *Trials* as presented in Table 4.4.

The above table shows that the sample average is the same for all the combinations of N and T, and it is indeed close to the *true mean* of 0.5. However, the sample variance changes (decreases) with the increasing number of experiments per trial (N), it is approximately equal to $\frac{S_x^2}{N}$ as is expected by the Central Limit Theorem!

4.7 GENERAL FORMULATION OF THE RELATIVE UNCERTAINTY

Here, for a random variable x with outcomes, x_i's, using Equation 4.37, we derive a general formulation for the relative uncertainty as follows

$$S_x^2 = \frac{1}{N-1} \sum_{i=1}^{N} (x_i - \overline{x})^2 = \frac{1}{N-1} \sum_{i=1}^{N} (x_i^2 + \overline{x}^2 - 2x_i\overline{x}) =$$

$$\frac{1}{N-1} \sum_{i=1}^{N} x_i^2 + \frac{1}{N-1} \sum_{i=1}^{N} \overline{x}^2 - \frac{2}{N-1} \overline{x} \sum_{i=1}^{N} x_i$$

$$S_x^2 = \frac{N}{N-1}\frac{1}{N}\sum_{i=1}^{N}x_i^2 + \frac{N}{N-1}\overline{x}^2 - \frac{2N}{N-1}\overline{x}\frac{1}{N}\sum_{i=1}^{N}x_i \qquad (4.120)$$

$$S_x^2 = \frac{N}{N-1}\overline{x^2} + \frac{N}{N-1}\overline{x}^2 - \frac{2N}{N-1}\overline{x}^2$$

$$S_x^2 = \frac{N}{N-1}(\overline{x^2} - \overline{x}^2)$$

Then, the relative uncertainty is given by

$$R_x = \frac{\sqrt{\frac{N}{N-1}(\overline{x^2}-\overline{x}^2)}}{\overline{x}} = \sqrt{\frac{N}{N-1}\left[\frac{\overline{x^2}}{\overline{x}^2}-1\right]} \qquad (4.121)$$

Now, if we consider that the Central Limit Theorem is valid, the above equation reduces to

$$R_{\overline{x}} = \frac{R_x}{\sqrt{N}} = \sqrt{\frac{1}{N-1}\left[\frac{\overline{x^2}}{\overline{x}^2}-1\right]} \qquad (4.122)$$

Now, if expand the average terms in the above equation, we obtain

$$R_{\overline{x}} = \sqrt{\frac{\frac{1}{N}\sum_{i=1}^{N}x_i^2}{(N-1)\left[\frac{1}{N}\sum_{i=1}^{N}x_i\right]^2} - \frac{1}{N-1}} \qquad (4.123)$$

Since $N \gg 1$, then we may replace $N-1$ by N, and the above equation reduces to

$$R_{\overline{x}} = \sqrt{\frac{\sum_{i=1}^{N}x_i^2}{\left[\sum_{i=1}^{N}x_i\right]^2} - \frac{1}{N}} \qquad (4.124)$$

Here, it is interesting to examine the use of Equation 4.124 for a Bernoulli random process. Considering that the outcomes of a Bernoulli process are either 1 or 0, then we can determine the summations in the Equation 4.124 as follows:

$$\sum_{i=1}^{N}x_i^2 = c$$

$$\left[\sum_{i=1}^{N}x_i\right]^2 = c^2$$

where c refers to number of counts (successes).

Using above values of the two summations in Equation 4.124, then the formulation for the relative uncertainty for a Bernoulli random variable reduces to

$$R_{\bar{x}} = \sqrt{\frac{c}{c^2} - \frac{1}{N}} = \sqrt{\frac{1}{c} - \frac{1}{N}} \qquad (4.125)$$

4.7.1 Special case of a Bernoulli random process

Here, it would be interesting to derive the formulation of the relative uncertainty by using the mean and variance of the Bernoulli process which were derived earlier.

Considering that after N sampling of a Bernoulli process, the number of successes is c, then the probability of success is equal to

$$p' = \frac{c}{N} \qquad (4.126)$$

Knowing that the true mean and variance of a Bernoulli process are p and pq, respectively, then we may derive the sample relative uncertainty as

$$R_x = \frac{S_x}{p'} = \frac{\sqrt{\frac{c}{N}\left(1 - \frac{c}{N}\right)}}{\frac{c}{N}} = \sqrt{\frac{N}{c} - 1} \qquad (4.127)$$

Now, considering that the Central Limit theorem is valid, the above equation reduces to

$$R_{\bar{x}} = \frac{R_x}{\sqrt{N}} = \sqrt{\frac{1}{c} - \frac{1}{N}} \qquad (4.128)$$

It is reassuring that Equation 4.128 agrees with Equation 4.125 that was derived from the general formulation of relative uncertainty, i.e., Equation 4.124.

4.8   CONFIDENCE LEVEL FOR FINITE SAMPLING

In this section, we discuss a method for estimating the confidence level in case of finite sampling. First, we will introduce the Student's t-distribution and then discuss its application for the estimation of the confidence level.

4.8.1 Student's t-distribution

The Central Limit Theorem states that for an infinite number of histories or experiments (N) per trial, the deviation of the average from the true mean, i.e., $t = \frac{\bar{x}-m_x}{\sigma_{\bar{x}}}$ follows a unit normal distribution. This behavior, however, may not happen for a finite value of N. The *Student t-distribution* was developed by William Gossett [95] to estimate a "better" variance for the finite N's. Gossett observed that for finite (or relatively small) N, generally a normal distribution overestimates the values near the mean and underestimates them away from the mean. So, he derived a distribution considering: (a) fitting the results of a sampling of finite N, and (b) including an adjustable parameter or a degree of freedom.

The probability density function for the Student's t-distribution with k *degrees of freedom* is expressed by

$$p_k(t) = \frac{\Gamma(\frac{k+1}{2})}{\sqrt{\pi k}\Gamma(\frac{k}{2})}\left(1+\frac{t^2}{k}\right)^{-\frac{k+1}{2}}, \qquad for \ -\infty < t < \infty, \ k = 1, 2, \cdots$$

(4.129)

The expected value of t is given by

$$m_t = E[t] = \int_{-\infty}^{\infty} dt t p_k(t) = 0 \tag{4.130}$$

The higher moments of t exist when the moment number (n) is less than the degree of freedom (k). The odd moments are equal to zero because of symmetry, and the even moments are given by

$$E[t^n] = \frac{k^n \Gamma(n+\frac{1}{2})\Gamma(\frac{k}{2}-n)}{\Gamma(\frac{1}{2})\Gamma(\frac{k}{2})}, \qquad for \ even \ n < k \tag{4.131}$$

The second moment or variance of a t-distribution is given by

$$\sigma_t^2 = E[t^2] - (E[t])^2 = \frac{k}{k-2} \tag{4.132}$$

As the number of degrees of freedom k goes to ∞, the t-distribution approaches the unit normal distribution, that is

$$lim_{k\to\infty}[p_k(t)] = \phi(t) = \frac{1}{\sqrt{2\pi}}e^{-\frac{t^2}{2}} \tag{4.133}$$

Note that in practical applications, when $k > 30$, the t-distribution is considered as a normal distribution.

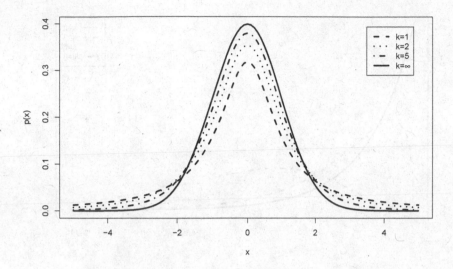

Figure 4.6 Comparison of t-distribution with normal distribution for different degrees of freedom

Figure 4.6 compares a t-distribution at different k-values to a normal distribution, i.e., $k \to \infty$. The figure indicates that the t-distribution has a lower value for the mean and higher values at the tails of the distribution.

Figure 4.7 examines the behavior of the chi-square distribution for different degrees of freedom. The figure indicates that a χ^2-distribution has a lower mean and relatively higher tails as compared to a normal density function.

Considering the above figures and the characteristic of a t-distribution, one can sample a t-distribution by sampling from a unit normal and χ^2-distribution using the following equation [83].

$$t = \frac{x}{\sqrt{\frac{\chi^2}{k}}} \tag{4.134}$$

where x and χ are independent random variables. The x random variable is sampled from a unit normal distribution, and the [w $=\chi^2$] random variable is sampled from a χ^2-distribution [$f_k(w)$] with k degrees of freedom.

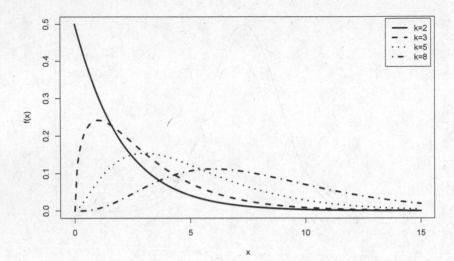

Figure 4.7 Behavior of the χ^2 distribution for different degrees of freedom

4.8.2 Determination of confidence level and application of the t-distribution

As mentioned earlier, in any Monte Carlo simulation, we need to determine the sample mean and variance, i.e., precision, and we have to attempt to estimate the confidence level on the relative uncertainty.

Generally, we do not know the true distribution, so we have to devise other means to determine the confidence level. For this, we may use the two limit theorems discussed in Section 4.5, i.e., the de Moivre–Laplace limit theorem and the central limit theorem. In the case of a finite sample size, there is a possibility that the distribution of the sample average is "seminormal," we, therefore, employ the Student's t-distribution to achieve a higher confidence.

Using a t-distribution with *N-1* degrees of freedom, the t-factor (t_{N-1}), which is equal to the number of standard deviations for achieving a 95% confidence level, is obtained. The t-factor is used to determine the 95% confidence interval (d_{95}) as follows

$$d_{95} = t_{N-1} S_{\overline{x}} \tag{4.135}$$

where $S_{\overline{x}} = \frac{S_x}{\sqrt{N}}$ is the estimated sample standard deviation for the sample average, $\overline{x}$.

The above confidence level then indicates that, for the given $\overline{x}$ and $S_{\overline{x}}$, there is 95% probability that the true mean (m_x) lies within the interval $[x - d_{95}, x + d_{95}]$.

4.9 TEST OF NORMALITY OF DISTRIBUTION

In addition to the use of the Student t-distribution to obtain a deviation interval of higher confidence, one can employ tests for examining the normality of the estimated sample average. This further improves confidence in the use of the central limit theorem. Here, we discuss two approaches including: (1) test of skewness coefficient and (2) w test.

4.9.1 Test of skewness coefficient

If the sample average is normally distributed, then the skewness coefficient has to be zero and, in the case of a finite sample size, should be smaller than an upper limit. The skewness coefficient C is determined by

$$C = \frac{1}{S_x^3} \frac{1}{N-1} \sum_{i=1}^{N} (x_i - \overline{x})^3 \tag{4.136}$$

If C is larger than an upper limit, then the user has to make a decision on the use of the derived confidence level. There are two situations that may arise

1. If $c^2 \geq (1.96)^2 \frac{6(N-1)(N-2)}{N(N+1)(N+3)}$, then the use of the confidence level is *doubtful*.

2. If $c^2 \geq (2.81)^2 \frac{6(N-1)(N-2)}{N(N+1)(N+3)}$, then the use of the confidence level is *meaningless*.

4.9.2 Shapiro-Wilk test for normality

The Shapiro–Wilk normality test [91], referred to as the *w-test*, determines a parameter w that should not exceed an expected value. The procedure for estimation of w is given below:

1. For the sample size of N with outcomes $(x_1, x_2, x_3, \cdots, x_N)$, compute the sample mean and sample variance using Equations 4.28 and 4.37, respectively.

2. Arrange x_i's in increasing order of magnitude, relabel them as y_i's, and then compute the following summation

$$b = \sum_{i=1}^{N} a_i y_i, \qquad (4.137)$$

where coefficients a_i are given in [91].

3. Compute w using the following formulation

$$w = \frac{b^2}{(N-1)S^2}. \qquad (4.138)$$

4. Using [91], for a given sample size (N) and w value, one can estimate the probability of normality of the sample average. If the w is small, e.g., < 0.806, this probability is low, $< 10\%$; while if $w > 0.986$, there is $> 95\%$ chance that the sample average follows a normal distribution.

In summary, both tests are straightforward, while the skewness test may require significantly less computing time.

PROBLEMS

1. Show that the number of permutations of n objects is equal to $n!$. (Note that permutation refers to a rearrangement of objects in some order.)

2. Show that the number of ordered samples of size k, with replacement, from n objects is n^k. (Ordered samples means that the order of the sampled elements matters, e.g., phone numbers, license plate, etc.)

3. Show that the number of ordered samples of size k, without replacement, from n objects is equal to:

$$P_n(k) = \frac{n!}{(n-k)!}$$

4. Show that the number of combinations of size k, without replacement, from n objects is equal to:

$$C_n(k) = \frac{n!}{k!\,(n-k)!}$$

5. Determine the number of license plates that are comprised of three letters and three digits.

6. Determine the number of phone numbers that are comprised of a three-digit area code, a three-digit local code, and a four-digit number.

7. A random process has an outcome x. If each outcome (x_i) is assigned a weight (w_i), then determine the weighted average of these outcomes and its associated standard deviation.

8. Prove that the variance of the "sample mean" is given by:

$$Var\,(\bar{x}) \equiv \sigma_{\bar{x}}^2 = E\left[(\bar{x}-m)^2\right] = \frac{\sigma_x^2}{N}$$

where:

$$\sigma_x^2 = E\left[(x-\bar{x})^2\right]$$

9. A random variable x has a density function $p(x) = x^2/9$ in a range $0 \le x \le 3$.

   a. Find the true mean of x.

   b. Find the true variance of x.

   c. Determine the expected value of $g(x) = 1/x$.

10. A random variable r has a probability density function $\Sigma e^{-\Sigma r}$ in a range $0 \le r \le \infty$:

   a. Find the true mean of r.

   b. Find the true variance of r.

   c. Find the expected value of $g(r) = 1/r$.

11. Random variable x has a density function expressed by:

$$f(x) = 1 + x, \quad \text{for} \quad 0 \le x \le 1$$

If function $g(x)$ is expressed by:

$$g(x) = x + cx^2,$$

determine the parameter c such that the variance of $g(x)$ is minimized.

12. Plot a Poisson distribution for a different average number of successes (m) including 1, 10, 50, 100, 1,000, and 10,000. Compare your results to a normal distribution with equivalent mean value.

13. It is known that the radioactivity decay process follows a Poisson distribution. For a radioactive nucleus, the remaining number of nuclei at time t is determined by:

$$n(t) = n_0 e^{-\lambda t}$$

where λ (decay constant) is referred to as the average number of disintegrations per second, therefore, the expected number of disintegrations (events) after time t is $m = \lambda t$.

a. Using Table 4.5 providing probabilities versus number of events for a given time t (generated based on the Poisson distribution for a radioactive nucleus), determine the average number of disintegrations.

Table 4.5 Probability table for Problem 13

Number of events	Probability	Num. of events	Probability
0	6.74×10^{-3}	9	3.63×10^{-2}
1	3.37×10^{-2}	10	1.81×10^{-2}
2	8.42×10^{-2}	11	8.24×10^{-3}
3	1.40×10^{-1}	12	3.43×10^{-3}
4	1.75×10^{-1}	13	1.32×10^{-3}
5	1.75×10^{-1}	14	4.72×10^{-4}
6	1.46×10^{-1}	15	1.57×10^{-4}
7	1.04×10^{-1}	16	4.91×10^{-5}
8	6.53×10^{-2}	17	1.45×10^{-5}

b. If the decay constant for a nucleus is $\lambda = 1.0 \ s^{-1}$, using the average number of disintegrations (decay events) from part (a), determine the decay time.

c. Determine the uncertainty associated with the number of disintegrated (decayed) nuclei based on the Poisson distribution and Bernoulli process.

14. A pollster wants to conduct a poll to determine the probable outcome of an election between candidates A and B. The pollster seeks 75% confidence that he knows the fraction of the vote ($f \pm$ 2%) that will go to candidate A. A preliminary poll has indicated that A will receive roughly 55% of the vote. How many voters should have been polled?

15. Write a program to sample the path length that a particle travels in a medium with $\Sigma = 2.0 \ cm^{-1}$. Considering that the path length is sampled from a probability density function given by $p(t) = e^{-r}$, estimate the sample mean and sample variance for the following combinations of trials and histories:

a. Total number of experiments, i.e., $trial \times (history/trial)$, is fixed to 1,000. Consider the following history/ trial ratios: 1, 2, 5, 10, 25, 50, 100, 200, and 500.

b. Number of history/trial ratio is fixed to 10. Consider the following number of trials: 10, 20, 30, 40, 50, 100, 200, and 400.

c. Number of trials is fixed to 20. Consider the following history/trial ratios: 1, 2, 5, 10, 25, 50, 100, 200, and 500.

Tabulate and plot your results and discuss your observations, especially regarding the Central Limit Theorem.

CHAPTER 5

Integrals and Associated Variance Reduction Techniques

CONTENTS

5.1 INTRODUCTION

One of the important applications of the Monte Carlo method is estimation of integrals of functions or of physical quantities. The method is highly flexible, and generally used for solving complex high-dimensional integrals or determining the outcomes of complex physical processes. The major issue is the need for significant computation time for achieving an acceptable precision or uncertainty. Therefore, over the past several decades, significant efforts have been allocated to the development of variance reduction approaches that result in smaller variance in a shorter computation time. Numerous articles and books from different scientific communities address techniques with varying success for different applications [29, 58, 66, 94, 40]

In this chapter, the fundamental concept of Monte Carlo integration is introduced, and several representative variance reduction methodologies often applied to particle transport problems are introduced. The performance of these methodologies is examined based on comparing their variance to that of the standard methodology. Through this presentation, it is expected that the reader will gain knowledge and understanding of variance reduction techniques, how to analyze their performance, and possibly how to explore the development of new innovative approaches.

5.2 EVALUATION OF INTEGRALS

The Monte Carlo method can be used to estimate a finite integral. This is realized if we inspect the definition of expected value of a random variable (x) with associated probability density function $(f(x))$ given by

$$I = E[x] = \int_a^b dx x f(x), \quad for \ x\epsilon[a,b], \tag{5.1}$$

or the expected value of an arbitrary function $g(x)$, defined in $[a,b]$, given by

$$I = E[x] = \int_a^b dx g(x) f(x), \quad for \ x\epsilon[a,b], \tag{5.2}$$

Hence, a finite integral is equivalent to the expected value of its integrand (or part of its integrand). For example, to evaluate the integrals in Equations 5.1 and 5.2, we may sample x from $f(x)$ and evaluate the average x as

$$I_N = \overline{x} = \frac{1}{N} \sum_{i=1}^N x_i \tag{5.3}$$

and the average $g(x)$ as

$$I_N = \overline{g(x)} = \frac{1}{N} \sum_{i=1}^N g(x_i) \tag{5.4}$$

In general, any integral can be evaluated via a Monte Carlo simulation after a *pdf* is identified for sampling the integration variable. For example, consider the following integral

$$I = \int_a^b dx h(x), \quad for \ x\epsilon[a,b] \tag{5.5}$$

Here, it is necessary to identify $g(x)$ (function being averaged) and $f(x)$ (*pdf*) such that $h(x) = f(x)g(x)$. The simplest thing is to sample (x) from a uniform distribution given by

$$f(x) = \frac{1}{b-a} \tag{5.6}$$

This implies that

$$g(x) = \frac{h(x)}{f(x)} = h(x)(b-a) \tag{5.7}$$

To sample $f(x)$, we form the corresponding FFMC expressed by

$$\int_a^x dx' \frac{1}{b-a} = \eta \tag{5.8}$$
$$x_i = a + \eta_i(b-a)$$

Then, the integral is calculated by

$$I_N = \overline{g(x)} = \frac{1}{N} \sum_{i=1}^N h(x_i)(b-a) \tag{5.9}$$

5.3 VARIANCE REDUCTION TECHNIQUES FOR DETERMINATION OF INTEGRALS

There are several techniques for variance reduction when determining integrals using the Monte Carlo method. These techniques are developed based on modification of different elements of an integral including *the pdf, the integrand, the domain of integration, and/or a combination of the aforementioned elements.* This section introduces *importance sampling, control variates, stratified sampling, and combined sampling techniques.*

Before discussing these techniques, it is important to introduce a metric for comparing the performance of different variance reduction techniques. A commonly used metric is the *figure of merit (FOM)* given by

$$FOM = \frac{1}{R_{\overline{x}}^2 T} \tag{5.10}$$

where $R_{\overline{x}}$ is the estimated relative uncertainty, and T is the computation time. Equation 5.10 indicates that a more effective technique should exhibit a higher FOM, i.e., the technique can achieve a smaller variance in a shorter time. It should be noted that the FOM for the same problem and same method will be different on different computers, i.e., it can only be used for relative comparison of different methods for the same problem on the same computer.

In the following discussions, we examine the effectiveness of each technique based on the reduction of variance only, as timing requires an implementation of each methodology that could be highly dependent on the application. Further discussion on FOM is given in Chapter 7.

5.3.1 Importance sampling

Importance sampling is based on modifying or changing the *pdf*, in order to achieve a low variance in a shorter time for a given tolerance. To identify the most effective *pdf*, we will introduce a new *pdf*, i.e., $f^*(x)$ by rewriting Equation 5.2 as follows

$$I = \int_a^b dx \left[\frac{g(x)f(x)}{f^*(x)} \right] f^*(x), \ for \ x\epsilon[a,b] \tag{5.11}$$

where,

$$f^*(x) \geq 0 \tag{5.12}$$

$$\int_a^b dx f^*(x) = 1, \ and \tag{5.13}$$

$$\frac{g(x)f(x)}{f^*(x)} < \infty, \ except \ at \ discrete \ set \ of \ points \tag{5.14}$$

In order to obtain a formulation for $f^*(x)$, we form the variance of the integrand and try to minimize the variance given by

$$Var[I] = \int_a^b dx \left[\frac{g(x)f(x)}{f^*(x)} \right]^2 f^*(x) - I^2 \tag{5.15}$$

Because the value of I is fixed irrespective of the selected $f^*(x)$, we have to minimize the first term in Equation 5.15, while maintaining the constraint expressed by Equation 5.13. To do so, we form the corresponding Lagrange multiplier expression given by

$$L[f^*(x)] = \int_a^b dx \frac{g^2(x)f^2(x)}{f^*(x)} + \lambda \int_a^b dx f^*(x) \tag{5.16}$$

and solve for $f^*(x)$ by minimizing the above equation with respect to $f^*(x)$, i.e.,

$$\frac{\partial L[f^*(x)]}{\partial f^*(x)} = 0 \tag{5.17}$$

To differentiate the integrals in Equation 5.16, we use the Leibnitz rule [3] expressed by

$$\frac{d}{d\alpha}\left[\int_{h(\alpha)}^{k(\alpha)} dx f(x, \alpha)\right] = \int_{h(\alpha)}^{k(\alpha)} d\alpha \frac{\partial f(x, \alpha)}{\partial \alpha} +$$
$$f(k(\alpha), \alpha)\frac{dk(\alpha)}{d\alpha} - f(h(\alpha), \alpha)\frac{dh(\alpha)}{d\alpha} \tag{5.18}$$

Using above formulation, Equation 5.17 reduces to

$$\frac{\partial L[f^*(x)]}{\partial f^*(x)} = -\int_a^b dx \frac{g^2(x)f^2(x)}{f^{*2}(x)} + \lambda \int_a^b dx = 0$$
$$\int_a^b dx \left[\frac{g^2(x)f^2(x)}{f^{*2}(x)} - \lambda\right] = 0 \tag{5.19}$$

To satisfy the above equality for any x, it is necessary that the integrand to be equal to zero, i.e.,

$$\frac{g^2(x)f^2(x)}{f^{*2}(x)} - \lambda = 0 \tag{5.20}$$

Hence, $f^*(x)$ formulation is given by

$$f^*(x) = \frac{|g(x)f(x)|}{\sqrt{\lambda}} \tag{5.21}$$

The value of λ can be determined by requiring that $f^*(x)$ is normalized, i.e., Equation 5.13,

$$\int_a^b dx \frac{g(x)f(x)}{\sqrt{\lambda}} = 1$$
$$\sqrt{\lambda} = I \tag{5.22}$$

Therefore, $f^*(x)$ is given by

$$f^*(x) = \frac{g(x)f(x)}{I} \tag{5.23}$$

Equation 5.23 cannot be used in practice because I is unknown; however, it indicates that the "best" pdf ($f^*(x)$) should be related to the integrand ($g(x)f(x)$).

Example 5.1

To demonstrate the use of the importance sampling formulation, i.e., Equation 5.23, let's consider calculating the following integral

$$I = \int_1^2 dx lnx$$

$$I = [xlnx - x]_1^2 = 0.386294 \tag{5.24}$$

A straightforward Monte Carlo simulation will sample (x) from a uniform distribution in the range of $[1,2]$. This means that the corresponding FFMC is expressed by

$$x = \eta + 1 \tag{5.25}$$

Therefore, the value of the integral after N samples using above equation is given by

$$I_N = \frac{1}{N} \sum_{i=1}^{N} \ln x_i \tag{5.26}$$

The analytical variance of the integral (I) is expressed by

$$Var[I] = \int_1^2 dx \ln^2 x - I^2$$

$$Var[I] = \left[(x \ln^2 x - 2x \ln x + 2x) - (x \ln x - x)^2 \right]_1^2 \tag{5.27}$$

$$Var[I] = 0.039094$$

To obtain a more effective *pdf* $(f^*(x))$, we expand $\ln x$ in a power series as

$$\ln x = \sum_{n=1}^{\infty} \frac{(-1)^{n+1}}{n} (x - 1)^n \tag{5.28}$$

For $n = 1$, $\ln x \simeq x - 1$, one may consider a more effective *pdf* as

$$f^*(x) = \alpha x. \tag{5.29}$$

Considering that $f^*(x)$ has to be normalized, α is determined by

$$\int_1^2 dx \alpha x = 1. \tag{5.30}$$

Solving above integral, results in

$$\alpha = \frac{2}{3}. \tag{5.31}$$

and therefore.

$$f^*(x) = \frac{2}{3}x \tag{5.32}$$

Considering above *pdf*, the integrand, $g^*(x)$, is expressed by

$$g^*(x) = \frac{\ln x}{\frac{2}{3}x} \tag{5.33}$$

Then, the integral (expected value of integrand) is determined by

$$I_N^* = \frac{1}{N} \sum_{i=1}^{N} \frac{\ln x_i}{\frac{2}{3}x_i} \tag{5.34}$$

where x_i is sampled from $f^*(x)$. The theoretical variance for the revised integrand is given by

$$Var[I] = \int_1^2 dx \left[\frac{\ln^2 x}{(\frac{2}{3}x)^2}\right]\left(\frac{2}{3}x\right) - I^2 \tag{5.35}$$

$$Var[I] = \left[0.5\ln^3 x\right]_1^2 = 0.017289$$

Hence, it is concluded that by considering only the first term of the $\ln x$ *expansion, we may achieve faster convergence by a factor of $\simeq 2$.*

5.3.2 Control variates technique

Again, consider the goal is to determine I expressed by Equation 5.2. Rather than modifying the pdf $(f(x))$, this time we explore changing the function $g(x)$ such that $Var[I]$ is reduced. As discussed by Kalos and Whitlock [58], one possibility is rewrite Equation 5.2 by subtracting and adding a function $(h(x))$ as follows

$$I = \int_a^b dx[g(x) - h(x)]f(x) + \int_a^b h(x)f(x) \tag{5.36}$$

To achieve a lower variance, $h(x)$ has to have the following conditions: (a) should be similar to $g(x)$, and (b) its weighted average, i.e., second integral has an analytical solution, I_h. This means that to evaluate I, we need to sample the first term only, i.e.,

$$I_N = \frac{1}{N} \sum_{i=1}^{N}[g(x_i) - h(x_i)] + I_h \tag{5.37}$$

If the above difference is almost a constant for all $x_i's$, or the term is almost proportional to $h(x)$, then it is expected that $Var[g(x)-h(x)] \leq Var[g(x)]$.

Example 5.2

To demonstrate this technique, let's consider the same example integral as the previous section, i.e.,

$$I = \int_1^2 dx \ln x \tag{5.38}$$

Then, let's consider $h(x) = x$, hence $Var[g(x) - h(x)]$ is given by

$$Var[g(x) - h(x)] = \int_1^2 dx(\ln x - x)^2 - \left[\int_1^2 dx(\ln x - x)\right]^2 \tag{5.39}$$

The first integral in the above equation is given by

$$\int_1^2 dx(\ln x - x)^2 =$$

$$\left[(x\ln^2 x - (2x - x^2)\ln x + 2x + \frac{x^2}{2} + \frac{x^3}{3})\right]_1^2 = 1.24906, \tag{5.40}$$

and the second integral is given by

$$\int_1^2 dx(\ln x - x)^2 = \left(\left[(x\ln x - x - \frac{x^3}{2})\right]_1^2\right)^2 = 1.24034.$$

Therefore,

$$Var[g(x) - h(x)] = 0.008722. \tag{5.41}$$

This means that, for this example, the *control variates* technique may result in a speedup of 4.5, which is twice as fast as importance sampling. In short, both techniques have resulted in speedup with relatively minimal changes to *pdf* and/or integrand.

5.3.3 Stratified sampling technique

This technique examines and partitions the domain of interest or integration, and sets the amount of sampling for each subdomain according to its importance or need for achieving acceptable reliability (precision). The technique is referred to as *stratification* and has been used in different applications.

One of the common applications of the *stratification* is the population survey area. Rather than doing a simple random sampling (SRS), one identifies "important" variables and establishes "homogeneous" subgroups (referred to as *strata*) based on these variables, then each subgroup is sampled according to its proportion in the population. If appropriate variables are selected, this method will have a higher precision over SRS because it focuses on important subgroups and avoids "unrepresentative" sampling. Stratification sampling also provides the ability to consider a smaller sample size. For example, for a population of size N partitioned into M subgroups with N_m population, if we consider a sample size of n ($\ll N$), the proportionate sample sizes of different subgroups (strata) can be calculated by

$$n_m = \left(\frac{N_m}{N}\right) n, \;\; m = 1, M \tag{5.42}$$

Since with use of the stratification sampling an appropriate number of samples is used for each stratum, it is possible to achieve a higher precision with a smaller sample size.

The other important application of this sampling approach is the evaluation of mathematical integrals or integral quantities. First, we discuss in detail the implementation process of the stratified sampling, then derive the formulation for selecting appropriate sample size resulting in reduced variance. As an example, again we consider integral I, Equation 5.2, and partition the integral domain $D \equiv [a, b]$ into M nonoverlapping subdomains (strata) $D_m{}'s$ such that

$$\int_D dx f(x) = \sum_{m=1}^{M} \int_{D_m} dx f(x) \tag{5.43}$$

Then, rewrite Equation 5.2 as

$$I = \sum_{m=1}^{M} \int_{D_m} dx f(x) g(x) \tag{5.44}$$

To proceed, first we define the *pdf* for sampling x for any subdomain (m) as

$$f_m(x) = \frac{f(x)}{\int_{D_m} dx f(x)} \tag{5.45}$$

and consider a formulation for the fractional area of each subdomain as

$$h_m = \frac{\int_{D_m} f(x)}{\int_D dx f(x)} = \int_{D_m} dx f(x) \qquad (5.46)$$

Then by combining Equations 11.4 and 5.46, $f(x)$ can be written as

$$f(x) = h_m f_m(x) \qquad (5.47)$$

Using above equation in Equation 5.44, integral I reduces to

$$I = \sum_{m=1}^{M} h_m \int_{D_m} dx f_m(x) g(x) \qquad (5.48)$$

Now, applying the Monte Carlo method for evaluating the subdomain integrals, the stratification formulation for integral (I_s) is given by

$$I_s = \sum_{m=1}^{M} h_m \frac{1}{N_m} \sum_{i=1}^{N_m} g_m(x_i) \qquad (5.49)$$

where, N_m is the number of samples obtained for subdomain m using the $f_m(x)$ probability density function. The ideal number of samples per subdomain (i.e., N_m) is not known; therefore, it is necessary to develop a formulation for its evaluation. To do so, we need to minimize the variance of I_s, while preserving the total number of samples, i.e., $N = \sum_{m=1}^{M} N_m$. The variance of I_s is given by

$$\sigma^2(I_s) = \sum_{m=1}^{M} \frac{h_m^2}{N_m^2} \sum_{i=1}^{N_m} \sigma^2(g_m(x_i))$$

$$\qquad (5.50)$$

$$\sigma^2(I_s) = \sum_{m=1}^{M} \frac{h_m^2}{N_m^2} N_m \sigma_m^2 = \sum_{m=1}^{M} \frac{h_m^2}{N_m} \sigma_m^2$$

where, $\sigma_m^2 \equiv \sigma(g_m(x))$. To minimize the variance, while maintaining a constraint on the total sample size, we use the *Lagrange multiplier* formulation expressed by

$$L[N_m] = \sum_{m=1}^{M} \frac{h_m^2}{N_m} \sigma_m^2 + \lambda \sum_{m=1}^{M} N_m \qquad (5.51)$$

Then, we minimize the *Lagrange multiplier* by finding its derivative with respect to N_m

$$\frac{\partial L[N_m]}{\partial N_m} = \sum_{m=1}^{M} \left[\frac{h_m^2}{N_m^2} \sigma_m^2 + \lambda \right] = 0 \tag{5.52}$$

To satisfy the above equality for any m, we set the bracket equal to zero, and obtain a formulation for λ in terms of N_m

$$\sqrt{\lambda} = \frac{h_m \sigma_m}{N_m}. \tag{5.53}$$

Then, we use the sample size constraint to obtain another formulation for $\sqrt{\lambda}$ as follows:

$$N = \sum_{m=1}^{M} N_m = \frac{1}{\sqrt{\lambda}} \sum_{m=1}^{M} h_m \sigma_m \tag{5.54}$$

therefore

$$\sqrt{\lambda} = \frac{\sum_{m=1}^{M} h_m \sigma_m}{N} \tag{5.55}$$

Then, if we equate Equations 5.53 and 5.55, we can derive a formulation for the sample size for each subdomain (m) as

$$N_m = \frac{h_m \sigma_m}{\sum_{m=1}^{M} h_m \sigma_m} N \tag{5.56}$$

Equation 5.56 indicates that the number of samples per subdomain is weighted by the combined effects of fractional area and variance of each subdomain. In other words, a subdomain with a larger combined area and variance requires more samples in order to obtain a minimum variance for a given total sample size N. Of course, this formulation is not useful unless one estimates the variances of the all the subdomains. This can be achieved by determining variances based on a relatively small number of sampling.

Now, if we substitute for N_m in Equation 5.50, then the variance formulation of I_s is expressed by

$$\sigma^2(I_s) = \sum_{m=1}^{M} \frac{h_m^2}{\frac{h_m \sigma_m}{\sum_{m=1}^{M} h_m \sigma_m} N} \sigma_m^2 \tag{5.57}$$

$$\sigma^2(I_s) = \frac{1}{N} \left[\sum_{m=1}^{M} h_m \sigma_m \right]^2 = \frac{1}{N} \overline{\sigma}_s^2$$

where, $\overline{\sigma}_s$ is given by

$$\overline{\sigma}_s = \sum_{m=1}^{M} h_m \sigma_m \tag{5.58}$$

To examine the effectiveness of the stratification approach, we have to compare above variance to the variance of the standard Monte Carlo formulation given by

$$I_N = \overline{g} = \frac{1}{N} \sum_{i=1}^{N} g(x_i) \tag{5.59}$$

The variance I_N is expressed by

$$\sigma^2(I_N) = \sigma^2 \left[\frac{1}{N} \sum_{i=1}^{N} g(x_i) \right] = \frac{1}{N^2} \sum_{i=1}^{N} \sigma^2(g(x_i)) = \frac{1}{N} \sigma_D^2 \tag{5.60}$$

where,

$$\sigma_D^2 = \int_D dx\, f(x)(g(x) - \overline{g})^2 \tag{5.61}$$

where, $\overline{g} = \int_D dx\, f(x)g(x)$. Using the stratified sampling technique, if we partition the domain (D) in Equation 5.61 into M subdomains, then $\sigma^2(I_N)$, Equation 5.60, reduces to

$$\sigma^2(I_N) = \frac{1}{N} \sum_{m=1}^{M} h_m \int_{D_m} dx\, f_m(x)(g(x) - \overline{g})^2 \tag{5.62}$$

Now, if we add and subtract $\overline{g}_m = \int_{D_m} dx\, f_m(x)g(x)$ in the parenthesis in Equation (5.61), the $\sigma^2(I_N)$ formulation reduces to

$$\sigma^2(I_N) = \frac{1}{N} \sum_{m=1}^{M} h_m \int_{D_m} dx\, f_m(x)(g(x) - \overline{g}_m)^2$$

$$+ \frac{1}{N} \sum_{m=1}^{M} h_m \int_{D_m} dx\, f_m(x)(\overline{g} - \overline{g}_m)^2 \tag{5.63}$$

$$+ \frac{2}{N} \sum_{m=1}^{M} h_m \int_{D_m} dx\, f_m(x)(g(x) - \overline{g}_m)(\overline{g}_m - \overline{g})$$

Above equation can be written as

$$\sigma^2(I_N) = \frac{1}{N} \sum_{m=1}^{M} h_m \sigma_m^2 + \frac{1}{N} \sum_{m=1}^{M} h_m (\bar{g} - \bar{g}_m)^2 \int_{D_m} dx f_m(x)$$

$$+ \frac{2}{N} \sum_{m=1}^{M} h_m (\bar{g}_m - \bar{g}) [\int_{D_m} dx f_m(x) g(x) - \bar{g}_m \int_{D_m} dx f_m(x)] \tag{5.64}$$

Considering that $\int_{D_m} dx f_m(x) = 1$ and $\int_{D_m} dx f_m(x) g(x) = \bar{g}_m$, above equation reduces to

$$\sigma^2(I_N) = \frac{1}{N} \sum_{m=1}^{M} h_m \sigma_m^2 + \frac{1}{N} \sum_{m=1}^{M} h_m (\bar{g} - \bar{g}_m)^2$$

$$+ \frac{2}{N} \sum_{m=1}^{M} h_m (\bar{g}_m - \bar{g}) [\bar{g}_m - \bar{g}_m)] \tag{5.65}$$

$$\sigma^2(I_N) = \frac{1}{N} \sum_{m=1}^{M} h_m \sigma_m^2 + \frac{1}{N} \sum_{m=1}^{M} h_m (\bar{g} - \bar{g}_m)^2$$

Now, to evaluate the effectiveness of the stratified sampling technique, the difference (Δ) between the standard variance ($\sigma^2(I_N)$), Equation 5.65, and the stratified sampling variance ($\sigma^2(I_s)$), Equation 5.50) is determined as follows

$$\Delta = \sigma^2(I_N) - \sigma^2(I_s) = \frac{1}{N} \sum_{m=1}^{M} h_m \sigma_m^2 + \frac{1}{N} \sum_{m=1}^{M} h_m (\bar{g} - \bar{g}_m)^2 - \frac{1}{N} \bar{\sigma}_s^2 \tag{5.66}$$

If we substitute Equation 5.58 for $\bar{\sigma}_s$, and combine first and third terms, above equation reduces to

$$\Delta = \frac{1}{N} \sum_{m=1}^{M} h_m (\sigma_m - \bar{\sigma}_s)^2 + \frac{1}{N} \sum_{m=1}^{M} h_m (\bar{g} - \bar{g}_m)^2 \tag{5.67}$$

Above formulation indicates that the difference (Δ) is always positive, and therefore

$$\sigma^2(I_N) > \sigma^2(I_s) \tag{5.68}$$

Equation 5.68 demonstrates that using an optimum set of $N_m's$, guarantees that the variance of a given integral when using the stratified sampling technique is always smaller than that achieved by the standard technique.

Example 5.3

To demonstrate the use of the stratified sampling technique, let's consider the following integral

$$I = \int_{-3}^{3} dx(1 + tanhx) \tag{5.69}$$

Given $g(x)f(x) = 1 + tanh(x)$, and considering a uniform probability density function

$$f(x) = \frac{1}{6}, \tag{5.70}$$

then Equation 5.69 reduces to

$$I = 6 \int_{-3}^{3} dx(1 + tanhx) \left(\frac{1}{6}\right). \tag{5.71}$$

Using Equations 11.4 and 5.46, we derive the $f_m(x)$ and h_m formulations, with D_m (subdomain interval) equal to 1, as follows

$$f_m(x) = \frac{\frac{1}{6}}{\int_{D_m} dx\frac{1}{6}} = 1$$

$$and \tag{5.72}$$

$$h_m = \int_{D_m} dx\frac{1}{6} = \frac{1}{6}$$

Using the previous information, formulations of variances, Equations 5.57 and 5.65, and Equation 5.56 for determination of an "optimum" number of samples, a computer program was written to estimate the integral, Equation 5.69, and its associated error for different total number of samples and different distributions of number of samples per interval. Three distributions of number of samples are considered including "standard" (obtained randomly from a uniform distribution), "uniform" (with equal number of samples per stratum), and "optimum" (obtained from Equation 5.56).

To generate an estimate of "optimum" distribution of number of samples per stratum, we first consider two cases with 100 and 1,000 total number of samples, divide the range of integral into 10 strata, and estimate the variance per stratum. Table 5.1 compares the estimated

Table 5.1 Comparison of number of samples per strata for the "standard" and "optimum" cases for 100 and 1000 sampled

Case	D1	D2	D3	D4	D5	D6	D7	D8	D9	D10
100 Samples										
standard	10	10	14	6	11	13	5	12	8	11
optimum	2	2	2	16	25	24	12	7	3	2
1000 samples										
standard	106	103	88	104	98	99	111	89	103	99
optimum	7	18	55	155	271	266	150	53	19	6

optimum number of samples per strata to those obtained using the standard approach for the two cases.

As expected the estimated number of samples (N_m) is highly variable in the 100-sample case as compared to the 1,000-sample case, which is approaching an expected uniform distribution. This, however, is not true for the optimum distribution, which has redistributed the number of samples based on the larger or smaller variance per strata. Figure 5.1 overlays the distribution of the integrand and the optimum distribution of number of samples.

Figure 5.1 indicates that the formulation for the optimum number of samples per strata (Equation 5.56) indeed has assigned significantly larger number of samples for the middle segment of integrand that exhibits large variation versus the two tails which exhibit smooth distributions.

To examine the effectiveness of the stratified sampling, we compare the analytical value of the integral (Equation 5.69), i.e., $I = 6.0$, to those estimated via the *standard* and the *stratified sampling* approaches. Table 5.2 compares the estimated value of the integral and associated standard deviation and relative difference (referred to 'rel') based on the different sampling approaches for increasing total number of samples between 100 and 100,000.

Table 5.2 indicates that the stratified sampling technique even with equal number of samples per strata yields significantly lower standard deviation and relative uncertainty as compared to the standard approach. As expected, the optimum stratified sampling performs the best. The estimated value of integral is in excellent agreement with the reference analytical value, especially for the optimum sampling

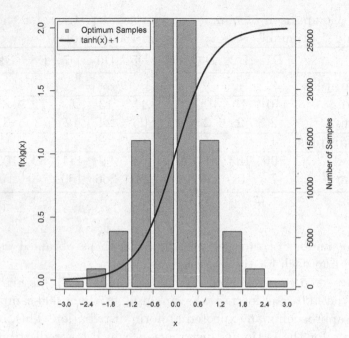

Figure 5.1 Distribution of integrand versus the optimum distribution of number of samples

approach that agrees within the estimated uncertainty. Figure 5.2 shows the expected behavior of the standard deviation for different sampling approaches with the number of samples as given in Table 5.2.

Theoretically, if the solution has converged to where the central limit theorem is valid, then the variance drops by the ratio of square roots of numbers of samples. For the current example, this ratio is $\frac{1}{\sqrt{10}}$ between two consecutive cases from 100 to 100,000.

To make sense of the effect of the uniform stratification, we start from Equation 5.65 for the integral variance of the standard sampling method, and add and subtract $(\frac{1}{N}\overline{\sigma}_s^2)$ to obtain

$$\sigma^2(I_N) = \frac{1}{N}\sum_{m=1}^{M} h_m(\sigma_m - \overline{\sigma}_s)^2 + \frac{1}{N}\sum_{m=1}^{M} h_m(\overline{g} - \overline{g}_m)^2 + \frac{1}{N}\overline{\sigma}_s^2 \quad (5.73)$$

Now, we analyze different terms in the above equation. With the uniform stratified sampling, the second term is eliminated. With optimum stratified sampling, the first term is also eliminated, leaving only

Table 5.2 Comparison of the predictions of the standard and stratified sampling techniques to the analytical solutions (Equation 5.69)

Method	Number of samples	Expected value	sample S_x	Rel. Diff. (%)
Standard				
	100	5.97237	0.50884	8.52
	1,000	5.94319	0.15494	2.61
	10,000	5.96752	0.04928	0.83
	100,000	5.99457	0.01555	0.26
Uniform stratified				
	100	5.97234	0.05588	0.94
	1,000	5.97917	0.01480	0.25
	10,000	5.99955	0.00492	0.08
	100,000	6.00088	0.00154	0.03
Optimum stratified				
	100	5.96951	0.03421	0.57
	1,000	6.01209	0.01104	0.18
	10,000	6.00326	0.00343	0.06
	100,000	6.00135	0.00109	0.02

the third term. In this example, the uniform stratified sampling performed almost as well as the optimum case, indicating that the second term was higher than the first term. The relative importance of these terms can vary significantly for different applications, and therefore impacting the effectiveness of the stratified sampling techniques.

5.3.4 Combined sampling

In principle, depending on the application, one can develop new techniques by combining the aforementioned variance reduction techniques. For example, one can combine the importance- and stratified-sampling approaches.

This means that after the $f^*(x)$ in Equation 5.11 is identified, and therefore, a new itegrand $g^\dagger(x)$ is formed, then the behavior of the integrand is considered for partitioning the domain into M subdomains.

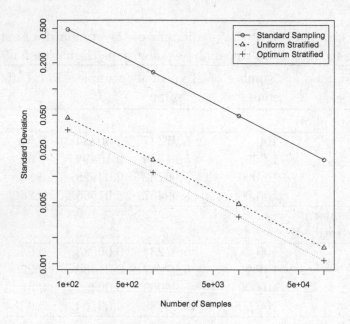

Figure 5.2 Variation of the standard deviation versus number of samples for different sampling approaches

This means that Equation 5.11 can be rewritten as

$$I = \sum_{m=1}^{M} h_m \int_{D_m} dx g^\dagger(x) f_m^*(x) \qquad (5.74)$$

where,

$$g^\dagger(x) = \frac{g(x)f(x)}{f^*(x)} \qquad (5.75)$$

and

$$f_m^*(x) = \frac{f^*(x)}{h_m} \qquad (5.76)$$

Then, for each subdomain, $f_m^*(x)$ is sampled to obtain the average value of integrand per subdomain as

$$\overline{g_m^\dagger} = \frac{1}{N_m} \sum_{i=1}^{N_m} g_m^\dagger(x_i) \qquad (5.77)$$

Hence, the integral is given by

$$I = \sum_{m=1}^{M} h_m \overline{g_m^\dagger} \qquad (5.78)$$

This method should yield a more accurate solution given that an appropriate *importance* function and sample size per subdomain are selected. Note that sample sizes have to be obtained based on Equation 5.56. An alternative approach to the above methodology is to use a unique *importance* function for each subdomain [103]. In addition to minimization of variance, it is important to make sure the computation time is also reduced, i.e., ultimately, the goal is to achieve high FOM.

5.4   REMARKS

In this chapter, we discussed how the Monte Carlo method can be used to estimate an integral, and how the variance of an integral can be reduced by modifying *pdf*, *integrand*, *sample size* of a partitioned set of subdomains, and/or using any combination of the aforementioned measures. It is demonstrated that, with relatively minimal effort, a significant reduction in variance can be expected. However, as mentioned, ultimately the FOM is the important parameter, because it accounts for the combined reduction of variance and computation time that is highly dependent on the application.

PROBLEMS

1. Determine the following integral using the importance sampling approach:

$$\int_0^1 dx \frac{1 + x + x^2}{1 + x}$$

Compare the efficiency to the standard sampling approach.

2. Use the importance sampling method to solve the integral:

$$\int_0^1 dx \, (x - sin(x))$$

Use the following functions to sample from:

   a. $f(x) = 1$.

   b. $f(x) = x$.

   c. An *intelligent* function of your choice.

For each function, sample until your relative error is less than 0.1%. Compare your computation times and number of samples needed to obtain the desired precision. Discuss your results.

3. Repeat Problem 2 (sampling from the same three functions), but change the range of integration to $[0, 10\pi]$. Compare your results to 1.

4. Compute the integral from Problem 1 using the correlation sampling technique.

5. Compute the integral from Problem 1 using stratified sampling. First, compute using five equal strata. Next, use 100 samples to estimate the optimum parameters before performing the rest of the samples.

6. Determine the integral:

$$\int_1^2 dx \, \ln(x)$$

using the standard sampling and uniform stratified sampling approaches. Examine the behavior of different terms in Equation 5.73.

7. Determine the integral:

$$\int_0^1 dx \, \sin(x)$$

using importance sampling, correlated sampling, and uniform stratified sampling. Compare the efficiency of the methods.

8. Determine the integral:

$$\int_{-3}^3 dx \, (1 + \tanh(x))$$

using importance sampling and correlated sampling.

9. Determine the integral:

$$\int_0^1 dx \frac{1}{1 + x^2}$$

using stratified sampling with standard, uniform, and optimum samples per stratum. Examine the behavior of the different terms in Equation 5.73.

CHAPTER 6

Fixed-Source Monte Carlo Particle Transport

CONTENTS

6.1   INTRODUCTION

This chapter is devoted to the application of the Monte Carlo method to a simplified particle transport problem. As indicated earlier, the method provides the ability of performing experiments on a computer and estimating expected behavior of particles and their interactions in a medium. For further discussion on the use of the Monte Carlo methods in particle transport, readers should consult the following books by Dunn and Shultis [29]; Greenspan, Kelber, and Okrent [41]; Kalos and Whitlock [58]; Lux and Koblinger [67]; Morin [77]; and Spanier and Gelbard [94]. Several code manuals, e.g., MCNP, PENLOPE, Serpent, also provide a good discussion on the methodology.

In general, a particle is emanated from a source (fixed or fission) with a random spatial position, random direction, and random energy. Each particle has a chance of traveling freely in a medium before undergoing an interaction with nuclei. Different types of interactions may occur, depending on the particle type and energy, and the composition of the medium. These interactions, which can be described by *pdfs* that are established using nuclear data and physics models, may lead to production of one or more particles, termination of the particle, and/or a change in particle energy or direction. A particle also may be terminated if it escapes from the medium. In a Monte Carlo simulation, the "history" of each particle from birth to death is followed, expected tallies of interest are estimated by simulation of numerous histories, and the associated variance and/or relative uncertainties for these tallies are evaluated.

In this chapter, we first introduce the linear Boltzmann equation and then describe the Monte Carlo method for neutron transport in a 1-D shield based on one-speed theory. This discussion includes a derivation of the necessary fundamental formulations of Monte Carlo (FFMCs) and an algorithm for developing a Monte Carlo code. In the last section, we introduce a correlated sampling methodology for performing perturbation studies.

6.2   INTRODUCTION TO THE LINEAR BOLTZMANN EQUATION

At this point, it is instructive to introduce the time-independent linear Boltzmann equation (LBE), which provides the particle balance in a phase space. Here, LBE is given for a nonmultiplying medium, but Chapter 11 is devoted to a detailed discussion on multiplying media or eigenvalue problems. Although our discussions are valid for any neutral particle transport, these examples of interactions refer to neutrons only. This is true throughout the rest of this book.

LBE expresses particle balance in a phase space $(d^3 r dE d\Omega)$. The time-independent LBE for a fixed-source is given by Bell and Glasstone [4].

$$\hat{\Omega} \cdot \vec{\nabla} \psi(\vec{r}, E, \hat{\Omega}) + \Sigma_t(\vec{r}, E) \psi(\vec{r}, E, \hat{\Omega}) =$$
$$\int_0^\infty dE' \int_{4\pi} d\Omega' \Sigma_s(\vec{r}, E' \to E, \hat{\Omega} \cdot \hat{\Omega}') \psi(\vec{r}, E', \hat{\Omega}') + S(\vec{r}, E, \hat{\Omega}),$$

$$(6.1)$$

Table 6.1 Description of terms in the LBE Equation 6.1

Term	Definition	Description
1	streaming	Expected particle streaming (flow) in a phase space
2	collision	Expected particle-nucleus interactions of any type in a phase space
3	scattering	Expected particle scattering from $(E', \hat{\Omega}')$ into $dEd\Omega$ about $(E, \hat{\Omega})$
4	source	Expected particle source density in a phase space

where $\psi(\vec{r}, E, \hat{\Omega}) = v(E)n(\vec{r}, E, \hat{\Omega})$ is the expected angular flux in phase space, where $v(E)$ is particle speed at energy E, and $n(\vec{r}, E, \hat{\Omega})$ is particle number density in phase space, $\Sigma_t(\vec{r}, E)$ is total macroscopic cross-section indicating probability per unit length of particle-nuclei interaction of all types at position $\vec{r}$ and energy E, $\Sigma(\vec{r}, E' \to E, \hat{\Omega} \cdot \hat{\Omega}')$ is the differential macroscopic scattering cross-section, indicating probability per unit length that following a particle-nuclei scattering a particle will scatter from energy E' to energy E within dE, and from direction (Ω') to direction (Ω) within $d\Omega$, and $S(\vec{r}, E, \hat{\Omega})$ is the fixed source density in phase space.

Physically, Equation 6.1 balances particle loss rate (left-hand side) and particle production rate (right-hand side) within a phase space. Specifically, each term is described in Table 6.1.

As stated in Table 6.1, the terms in Equation 6.1 represent expected (average) values of particle loss and production through different random processes. The Monte Carlo method can be used by averaging histories (from birth to death) of individual particles to calculate different terms in Equation 6.1.

Because the majority of Monte Carlo concepts and techniques can be discussed without the need for multidimensionality and energy dependency, most of our discussions in this book are based on 1-D, one-speed particle transport, with isotropic elastic-scattering and source, expressed by

$$\mu \frac{\partial \psi(x, \mu)}{\partial x} + \Sigma_t(x)\psi(x, \mu) = \frac{1}{2}(\Sigma_s \phi(x) + S(x, \mu)), \qquad (6.2)$$

where x refers to the dimension in which physical properties are changing, μ refers to the direction cosine of the particle direction $(\hat{\Omega})$ relative to the $x-axis$, and $\phi(x)$ refers to the scalar flux obtained by integrating the angular flux (i.e., $\int_{-1}^{1} d\mu\psi(x,\mu)$). For further simplicity, the source is considered to be generated at a point (x_0) and along one direction (μ_n), i.e.,

$$S(x,\mu) = S_0\delta(x - x_0)\delta(\mu - \mu_n) \qquad (6.3)$$

Only two types of interactions, scattering and absorption, are considered here. This means that $\Sigma_t = \Sigma_a + \Sigma_a$, and absorption is considered synonymous with *capture*, which terminates a particle history. For completeness, discussions on different types of interactions and sampling formulations for energy dependence as well as elastic and inelastic scattering are presented in Appendix 5.

6.3 MONTE CARLO METHOD FOR SIMPLIFIED PARTICLE TRANSPORT

Consider a monodirectional, one-speed neutron point source placed at the left boundary of a 1-D homogeneous shield, which is placed in a vacuum, with a detector at its right boundary, as shown in Figure 6.1.

In order to determine the neutron distribution within or beyond the shield, i.e., at the detector, or the rate of transmission or reflection of particles through the shield via the Monte Carlo method, first we need to identify the basic physical processes, and to obtain their associated *pdf s*. Further, we need to derive the corresponding FFMCs for sampling random variables corresponding to different random processes. Note

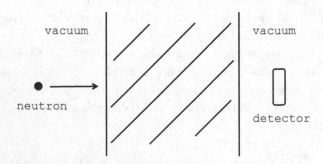

Figure 6.1 Schematic of a 1-D shield

that since the shield is placed in a vacuum, any particle that leaves the shield will not return, i.e., nonreentrant boundary condition.

In this simplified particle transport, there are only three basic random processes as follows:

1. Path-length (or *free flight*)

2. Interaction type (absorption or scattering)

3. Scattering angle in the case of scattering interaction; only "elastic" scattering is considered

6.3.1 Sampling path length

Here, we have to derive the combined probability of a particle moving freely a distance r, followed by probability of interaction in dr. From basic particle physics, we know that:

1. probability of free travel in a distance r is $e^{-\Sigma_t r}$

2. probability of an interaction in a distance dr is $\Sigma_t dr$.

Hence, the combined probability of above processes is given by

$$p(r)dr = \Sigma_t e^{-\Sigma_t r} dr \qquad (6.4)$$

Now, we form the FFMC for this process as

$$P(r) \equiv \int_0^r dr' \Sigma_t e^{-\Sigma_t r'} = \eta, \qquad (6.5)$$

and solve the path length as follows

$$1 - e^{-\Sigma_t r} = \eta$$
$$\Sigma_t r = -\ln(1 - \eta) \qquad (6.6)$$
$$r = -\frac{\ln \eta}{\Sigma_t}$$

Note that to reduce arithmetic operations in Equation 6.6, rather than using $1 - \eta$ sequence of random numbers, we use simply η sequence of random numbers.

For a multiregion problem, Σ_t may change from one region to other, therefore, using Equation 6.6 can be highly inefficient, as every time particle crosses an interface of differnt materials, one has bring back the particle to the interface and sample the path length using the correct Σ_t. In such situations, one may sample the distance (r) in terms of the number mean-free-paths (mfps, λ). The mfp which is the expected (average) distance that a particle travels freely is calculated by

$$\lambda \equiv E(r) = \int_0^\infty dr(r)(\Sigma_t e^{-\Sigma_t r}) \tag{6.7}$$

using the integration-by-parts approach, above equation reduces to

$$\lambda = -re^{-\Sigma_t r}|_0^\infty - \int_0^\infty dre^{-\Sigma_t r} = 0 - \frac{1}{\Sigma_t}e^{-\Sigma_t r}|_0^\infty$$
$$\lambda = \frac{1}{\Sigma_t} \tag{6.8}$$

Now, we rewrite Equation 6.6 in terms of mfp as

$$r = -\frac{\ln \eta}{\frac{1}{\lambda}}$$
$$b \equiv \frac{r}{\lambda} = -\ln \eta \tag{6.9}$$

where b refers to the number of mfp's. The procedure for the use of the number of $mfp's$ approach is follows:

1. determine $b_m = \Sigma_{i=1}^m \Sigma_{t,i} r_i$ for each subregion (m) along the direction of motion of particle as depicted in Figure 6.2.

2. generate a random number (η), march from region $\hat{m}$ (particle position) toward the last subregion along the particle direction until the following inequality is satisfied, indicating that the particle is in region m.

$$b_{m-1} < -\ln \eta \leq b_m \tag{6.10}$$

3. calculate r within m^{th} region as

$$r = -\frac{\ln \eta + b_{m-1}}{\Sigma_{t,m}} \tag{6.11}$$

It is worth noting that the number of mfp approach become more beneficial with increasing number of regions, especially if $\Sigma_{t,m}$ is very small for all subregions.

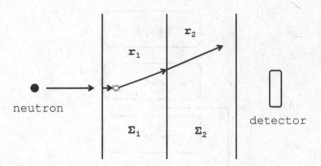

Figure 6.2 Schematic of a 1-D shield

6.3.2 Sampling interaction type

Following its free flight, the particle undergoes an interaction. The question is: What type of interaction? As mentioned, we are considering only absorption (capture) and "elastic" scattering. This is a binary random process and the probability of absorption is $(p_a = \frac{\Sigma_a}{\Sigma_t})$, the probability of scattering is $(p_s = 1 - \frac{\Sigma_a}{\Sigma_t})$. This means that $\frac{\Sigma_a}{\Sigma_t}$ gives the fraction of absorption interactions and the remainder are scattering interactions. The procedure for this binary set of interactions is given below:

1. generate a random number (η)

2. If $\eta \leq \frac{\Sigma_a}{\Sigma_t}$, the particle is absorbed, otherwise it is scattered.

Note that above procedure follows Equation (2.16) for discrete random variables, i.e., if $\eta \leq min[P_1, P_2] = min[\frac{\Sigma_a}{\Sigma_t}, 1]$, then particle is absorbed; otherwise, it is scattered.

6.3.2.1 Procedure for $N(> 2)$ interaction type

In general, if there are N different interactions, we consider the following steps:

1. Determine the interaction probabilities $(p_n's)$ corresponding to different interaction types, i.e., $n = 1, N$.

2. For any arbitrary list of interaction types, determine the corresponding CFD's, $P_n = \sum_{n'=1}^{n} p_{n'}$.

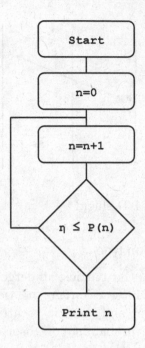

Figure 6.3 A flowchart for sampling a discrete random variable with N outcomes

3. Generate a random number (η), and if η satisfy the following inequality, then the outcome is interaction type n.

$$P_{n-1} < \eta \leq P_n \tag{6.12}$$

Figure 6.3 shows a flowchart for developing a software for Equation 6.12.

6.3.2.2 Procedure for a discrete random variable with N outcomes of equal probabilities

For a special case in which all interactions have an equal chance of occurrence, i.e., distribution; $p_n = \frac{1}{N}$, this flowchart can be simplified as follows:

- For a discrete a random variable, the *cdf*, P_n has to be related to η

- Hence, $\frac{n}{N} \propto \eta$, or $n \propto N \cdot \eta$

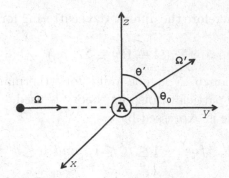

Figure 6.4 Schematic of a scattering process

- Since interaction type is an integer number, then we have to consider $n \propto INT(N \cdot \eta)$

- Then, to be able sample interaction types 1 and N, above can be reduced to the following equality:

$$n = INT(N \cdot \eta) + 1 \tag{6.13}$$

Of course, Equation 6.13 provides a highly efficient method for sampling a discrete random variable with N outcomes of equal probabilities.

6.3.3 Selection of scattering angle

If the interaction type is scattering, we need to determine the scattering angle. To do so, one has to derive a *pdf* for scattering angle distribution. From one-speed particle transport theory, the differential scattering cross-section is expressed by

$$\Sigma_s(\hat{\Omega} \to \hat{\Omega}')d\Omega' \equiv \tag{6.14}$$

probability per unit length that a particle moving along direction $\hat{\Omega}$ will scatter into a solid angle $(d\Omega')$ about direction Ω'. (Here, the unit of the differential scattering cross-section is $\left(\frac{1}{cm-steradian}\right)$.

Figure 6.4 depicts definition of the differential scattering formulation in a 3-D (x, y, z) coordinate system. For a discussion on the formulation for solid angle, please refer to Appendix 3.

Considering that in the one-speed transport theory [4], the differential scattering is dependent (only) on the scattering angle, i.e.,

$\mu_0 = \Omega \cdot \Omega'$, and therefore the differential scattering formulation reduces to

$$\Sigma_s(\hat{\Omega} \to \hat{\Omega}') \equiv \Sigma_s(\mu_0) \tag{6.15}$$

Note that Equation 6.15 is not valid for "thermal" energies when energy-dependent particle transport theory is used.

The solid angle is expressed by

$$d\Omega' = d\mu' d\phi', \quad for \ -1 \le \mu' \le 1, \ and \ 0 \le \phi' \le 2\pi, \tag{6.16}$$

where $\mu = \hat{i} \cdot \hat{\Omega}' = cos(\theta')$, and θ' and ϕ' are referred to as *polar* angle and *azimuthal* angle, respectively. Here, $\hat{i}$ referred to unit vector along $x - axis$ in a 3-D frame of reference.

Since the differential scattering process is comprised of two random variables, i.e., scattering interaction and direction after scattering (scattering angle), then we may write the following equality

$$\Sigma_s(\mu_0)d\mu'd\phi' = \Sigma_s p'(\mu_0)d\mu'd\phi', \tag{6.17}$$

where Σ_s is scattering cross-section (i.e., probability per unit length for scattering) and $p'(\mu_0)$ is probability per steradian that the scattered particle will change direction to new direction Ω' within the solid angle $d\Omega'$.

To proceed further, it is necessary to find a relation between μ' and μ_0, and rewrite above equation. To avoid mathematical complication, here first we perform a rotation of system of coordinates such that $z - axis$ become aligned with the particle direction before scattering (i.e., Ω). Then, as shown in Figure 6.5, $\mu' = \mu_0$ and $\phi' = \phi_0$, and therefore Equation 6.17 reduces to

$$\Sigma_s(\mu_0)d\mu_0 d\phi_0 = \Sigma_s p'(\mu_0)d\mu_0 d\phi_0 \tag{6.18}$$

Note that normally after a rotation of the system of coordinates, it is necessary modify equations to account for rotation effects, but this is not necessary here as the variable of interest is μ_0 that does not change, and therefore, $\Sigma_s(\mu_0)$ and $p(\mu_0)$ are not affected.

Next, we integrate Equation 6.18 over $\phi_0 \epsilon [0, 2\pi]$, and solve for $p(\mu_0)$ as follows

$$p(\mu_0)d\mu_0 = \frac{\int_0^{2\pi} d\phi_0 \Sigma_s(\mu_0)d\mu_0}{\Sigma_s}$$

$$p(\mu_0)d\mu_0 = 2\pi \frac{\Sigma_s(\mu_0)}{\Sigma_s}d\mu_0 \tag{6.19}$$

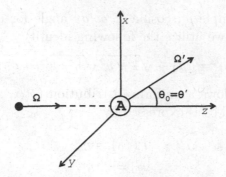

Figure 6.5 A flowchart for sampling a discrete random variable with N outcomes

Note that we have chosen to set $p(\mu_0) = \int_0^{2\pi} d\phi_0 p'(\mu_0)$, while performing integration on the right-hand side of above equation. This measure maintains that the unit of differential cross-section remains per unit steradian., as the available data commonly are given per unit steradian.

Now, if we consider scattering process is isotropic, i.e., $\Sigma_s(\mu_0)$ is equal to a constant c, then we have to derive the value of by considering that

$$\int_0^{2\pi} d\phi_0 \int_{-1}^{1} d\mu_0 \Sigma_s(\mu_0) = \Sigma_s$$

$$2\pi \int_{-1}^{1} d\mu_0 c = \Sigma_s \tag{6.20}$$

$$c = \frac{\Sigma_s}{4\pi}$$

Therefore, the differential scattering formulation for an isotropic scattering is expressed by

$$p(\mu_0) = \frac{1}{4\pi} \tag{6.21}$$

Now, we derive the FFMC for sampling the scattering angle (μ_0) as follows

$$2\pi \int_{-1}^{\mu_0} d\mu' (\frac{\Sigma_s}{4\pi}) = \eta \tag{6.22}$$

Solving above equation results in an for sampling an isotropic scattering process:

$$\mu_0 = 2\eta - 1 \tag{6.23}$$

Now, we have sample μ', cosine of *polar* angle for direction Ω' after scattering. To so, we utilize the following identity:

$$\mu' = \mu\mu_0 + \sqrt{1 - \mu^2}\sqrt{1 - \mu'^2}cos\phi_0 \tag{6.24}$$

Where, ϕ_0 follows a uniform distribution $(p(\phi_0) = \frac{1}{2\pi}$, hence the corresponding FFMC is expressed by

$$\begin{aligned} P(\phi_0) = \eta \\ \phi_0 = 2\pi\eta \end{aligned} \tag{6.25}$$

In a 3-D domain, it is necessary to sample direction cosines along the three axes $(x, y, \& z)$. Hence, $\hat{\Omega}$ and $\hat{\Omega}'$ are expressed as

$$\hat{\Omega} = u\hat{i} + v\hat{j} + w\hat{k},$$

and $\tag{6.26}$

$$\hat{\Omega}' = u'\hat{i} + v'\hat{j} + w'\hat{k}$$

Considering that in 3-D, the scattering angle identified by its *polar* angle (θ_0) and *azimuthal* angle (ϕ_0), then the direction cosines for particle direction after scattering are obtained via the following equations:

$$u' = \hat{i} \cdot \hat{\Omega}' = -(\frac{uw}{s} - \frac{v}{s})sin\theta_0 + ucos\theta_0 \tag{6.27}$$

$$v' = \hat{j} \cdot \hat{\Omega}' = -(\frac{vw}{s} - \frac{u}{s})sin\theta_0 + vcos\theta_0 \tag{6.28}$$

$$w' = \hat{k} \cdot \hat{\Omega}' = (s)(sin\theta_0)(cos\phi_0) + wcos\theta_0 \tag{6.29}$$

where $s = \sqrt{1 - w^2}$. A detailed discussion on the derivation of the above equations is given in Appendix A3.

6.4 A 1-D MONTE CARLO ALGORITHM

Based on the discussions in the previous section, a flowchart for Monte Carlo particle transport in a 1-D shield, considering one-speed theory, scattering and capture interactions, and isotropic scattering, is presented in Figure 6.6. Note that in the flowchart, there is a reference to particle weight (w) that will be discussed in Chapters 7 and 8. For the current discussion, the particle weight has to be set to 1.

In Figure 6.6, first the problem is initialized by defining parameters, including the maximum number of particles(n_{max}), shield thickness (L), relative uncertainty tolerance (tol), etc. Next, the particle

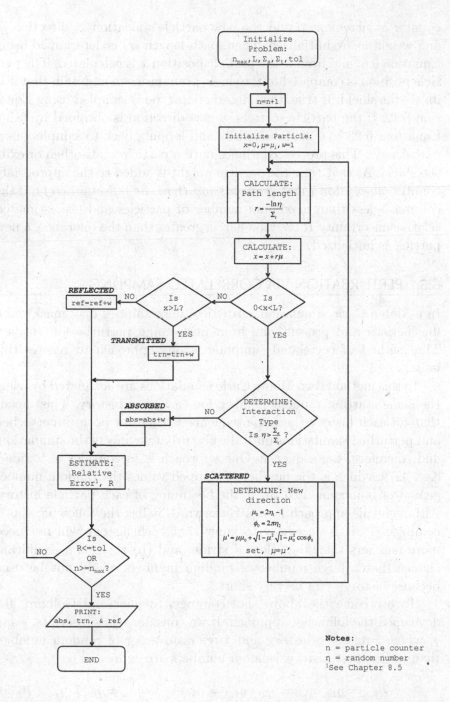

Figure 6.6 A flowchart for a Monte Carlo algorithm for simulation of a 1-D shield

counter is incremented and the new particle's location x, direction μ, and weight w are initialized. Then, path-length (r) is determined using Equation 6.6, and from it the particle position x is calculated. The particle position is compared to the shield boundaries to maintain that it is inside the shield. If true, then, interaction type is sampled using Equation 6.12. If the particle scatters, a new direction is obtained by using Equations 6.23 to 6.25, followed by and looping back to sample a new path-length. This process continues until a particle is absorbed or exits the shield. At that time the particle weight is added to the appropriate counter: *absorption (abs), transmission (trn), or reflection (ref)*. If the counter is less than maximum number of particles and the estimated relative uncertainty R (Section 8.5) is greater than the tolerance, a new particle is initialized.

6.5 PERTURBATION VIA CORRELATED SAMPLING

In a Monte Carlo simulation, statistical uncertainties may mask small fluctuations and prevent one from performing perturbation studies. The method of correlated sampling is one approach to resolve this issue.

In this method, two Monte Carlo simulations are correlated by using the same starting random number for each new history. This means that for each history the same sequence is used for both unperturbed and perturbed simulations until the perturbation affects the simulation, and, therefore, the sequence. One approach is to consider a "stride," e.g., 4279, which is the number of skipped values the random number generator is incremented by at the beginning of each particle history. Although this approach is straightforward, it has the following shortcomings: (a) it is necessary to know that each history will not need more numbers than the selected stride, and (b) there is a significant chance that a large number of random numbers would not be used because histories can be very short.

To overcome the above shortcomings, Spanier and Gelbard [94] developed the following approach. If we consider two subscripts, i and j, where i refers to history and j refers to order of random number, then for the first history, random numbers are expressed as

$$\eta_{1,1} = \eta_1, \ \ \eta_{1,2} = \eta_2, \ \ \eta_{1,3} = \eta_3, \cdots \cdots \eta_{1,n} = \eta_n \qquad (6.30)$$

For the second through N^{th} histories, every history is started by using the complement of the second random number from the previous history as follows

$$\eta_{2,1} = 1 - \eta_{1,2}, \quad \eta_{2,2} = PRNG(\eta_{2,1}, 1), \quad \eta_{2,3} = PRNG(\eta_{2,1}, 2), \cdots\cdots\cdots$$
$$\cdots\cdots \eta_{2,n} = PRNG(\eta_{2,1}, n-1)$$
$$\eta_{3,1} = 1 - \eta_{2,2}, \quad \eta_{3,2} = PRNG(\eta_{3,1}, 1), \quad \eta_{3,3} = PRNG(\eta_{3,1}, 2), \cdots\cdots\cdots$$
$$\cdots\cdots \eta_{3,n} = PRNG(\eta_{3,1}, n-1)$$
$$\eta_{N,1} = 1 - \eta_{N-1,2}, \quad \eta_{N,2} = PRNG(\eta_{N,1}, 1), \quad \eta_{N,3} = PRNG(\eta_{N,1}, 2), \cdots$$
$$\cdots\cdots \eta_{N,,n} = PRNG(\eta_{N,1}, n-1)$$
$$(6.31)$$

where $PRNG(\eta_{n,1}, k)$ refers to a pseudo random number generator that uses seed $\eta_{n,1}$ in the nth history to generate k^{th} random number in the sequence. Note that in this approach, Equations 6.30 and 6.31, for each history, at least two random numbers have to be generated even if the history is terminated after the use of the first random number.

6.6 HOW TO EXAMINE STATISTICAL RELIABILITY OF MONTE CARLO RESULTS

As discussed in Chapter 4, the results of a Monte Carlo simulation should include average quantities and their associated uncertainties. Further, a Monte Carlo simulation generally is terminated after a prescribed precision is achieved, i.e., when the relative uncertainty is less than a given tolerance. Considering that the number of particle histories, N, is *very large*, the variance corresponding to the sample average is calculated by applying the *central limit theorem (CLT)*, i.e., $\sigma_{\bar{x}}^2 = \frac{\sigma_x^2}{N}$. Since it is not possible to guess if N is sufficiently *large* for a given problem, one should at least examine the behavior of *relatve uncertainty* and *FOM*.

To examine the *relative uncertainty*, we determine the ratio of relative uncertainties (R_1 and R_2) for increasing number of histories, from N_1 to N_2, as follows.

$$\frac{R_2}{R_1} = \frac{\frac{\sigma_{x,2}}{\bar{x}_2 \sqrt{N_2}}}{\frac{\sigma_{x,1}}{\bar{x}_1 \sqrt{N_1}}} \qquad (6.32)$$

If CLT is valid then $\frac{\sigma_{x,1}}{\overline{x}_1} = \frac{\sigma_{x,2}}{\overline{x}_2}$ therefore above ratio reduces to

$$\frac{R_2}{R_1} = \sqrt{\frac{N_1}{N_2}} \qquad (6.33)$$

Hence, a reliable result of a Monte Carlo simulation should follow Equation 6.33.

For FOM, if CTL is valid, then

$$R_{\overline{x}} \simeq \frac{\sigma_x}{\sqrt{N}} \simeq \frac{C_1}{\sqrt{N}}, \qquad (6.34)$$

and considering computer time (T) should change with the number of histories, i.e., $T \simeq c_2 N$, then the FOM reduces to

$$FOM \simeq \frac{1}{(\frac{c_1}{\sqrt{N}})^2 (C_2 N)} \simeq C \qquad (6.35)$$

This means that for a reliable Monte Carlo simulation, it is necessary that the corresponding FOM fluctuates about a constant value (C).

6.7 REMARKS

This chapter provides an introduction to Monte Carlo particle transport. For a simple 1-D, one-speed neutron shield problem, it identifies the three random processes. Using the FFMC, it derives the necessary formulations for sampling each random variable and elaborates on a Monte Carlo simulation for a 1-D shield. It provides an algorithm for 1-D, one-speed Monte Carlo neutron transport simulation. A methodology for performing perturbation studies is presented. Finally, a discussion on analyzing the reliability of the results is provided. In conclusion, any result from a Monte Carlo simulation should be carefully analyzed, otherwise it is easy to achieve a *precise inaccurate* result.

PROBLEMS

1. The distribution function for a continuous random variable r is given by:

$$f(r) = e^{-r}, \qquad \text{for} \quad r \in [0, 8]$$

Derive the corresponding FFMC.

2. The distribution function for a continuous random variable x is given by:

$$f(x) = 1 + x^2 + x^3, \qquad \text{for} \quad x \in [0, 1]$$

Derive the corresponding FFMC.

3. The probability density function of the scattering angle is expressed by:

$$p(\mu_0) = 2\pi \frac{\Sigma_s(\mu_0)}{\Sigma_s}$$

a. If $\Sigma_s(\mu_0) = k$, derive the value of k in terms of scattering cross-section Σ_s.

b. If $\Sigma_s(\mu_0) = a\mu_0^2$, then derive the FFMC formulation for sampling the scattering angle.

4. Write a program for estimating the probability of transmission, reflection, and absorption of particles passing through a 1-D, multiregion shield with an arbitrary number of regions. Transmission is defined as a particle passing through to the other side of the shield. Reflection is defined as a particle exiting the shield from the starting side. Consider that source particles are emitted normal to the surface. For sampling the distance that a particle moves freely between collisions, implement two algorithms based on the *path-length* and *number of mean-free-path* approaches.

a. Derive analytical expression for a purely absorbing shield of $\Sigma_t = 1.0 \frac{1}{cm}$, and thickness $= 5.0 cm$.

b. Examine the accuracy of your code based on the purely absorbing shield in part (a).

c. Examine your code for a shield with 10% isotropic scattering.

5. Examine the performance of your code from Problem 4 for the following 3-region shield:

a. Case 1:
   Region 1: $\Sigma_t = 0.1 \ cm^{-1}$, $\Sigma_a = 0.01 \ cm^{-1}$, thickness $= 0.10 \ cm$
   Region 2: $\Sigma_t = 10 \ cm^{-1}$, $\Sigma_a = 0.1 \ cm^{-1}$, thickness $= 0.10 \ cm$

Table 6.2 Region properties for Problem 6.

Region	Σ_t cm^{-1}	Σ_a cm^{-1}	Thickness (cm)
1	0.5	0.4	1.0
2	0.1	0.03	0.10
3	0.2	0.2	2.0

Region 3: $\Sigma_t = 100 \ cm^{-1}$, $\Sigma_a = 10 \ cm^{-1}$, thickness $= 0.10 \ cm$

b. Case 2: same as (a), except switch regions 1 and 3.

c. Case 3: same as (a), except remove region 3.

Run each case for 1,000, 100,000, and 10,000,000 starting particle histories. Give the calculated probabilities of transmission, reflection, and absorption and associated uncertainties for each sampling method, i.e., *path-length* and *number of mfp*, each case and each number of particle histories. Discuss your results. You may use the flowchart given in Figure 6.6 to help with this program.

6. Use the program developed in Problem 4 to examine the correlated method used for testing a small perturbation in material properties of one region. The test case is comprised of three regions as presented in Table 6.2.

a Increase absorption fraction of region 2 by 5%, 10%, and 50%, and determine the impact on different probabilities.

b Decrease the thickness of region 2 by 5%, 10%, and 50%, and determine the impact on different probabilities.

c Compare and discuss your results in parts (a) and (b).

7. Using your program from Problem 4, analyze three 6-region shields with 6 different materials as given in Table 6.3. Examine the performance of the *path-length* and *number of mfp* approaches for these shields. Each region is 1 cm wide. Note that $c = \frac{\Sigma_s}{\Sigma_t}$.

Table 6.3 Region properties for Problem 7.

Case	Region 1	Regions 2-6
1	$\Sigma_t = 1\ cm^{-1}$ $c = 50\%$	Incremental reduction of Σ_t by 10%, i.e., the 6th region has a $\Sigma_t = 0.5\ cm^{-1}$
2	$\Sigma_t = 1\ cm^{-1}$ $c = 70\%$	(same as case 1)
3	$\Sigma_t = 1\ cm^{-1}$ $c = 90\%$	(same as case 1)

8. Modify the algorithm you developed in Problem 4 by considering that the scattering process is expressed by:

$$p\left(\mu_0\right) = \frac{1}{2}\left(1 + \mu_0\right)$$

Examine the algorithm based on the shields given in Problem 5.

CHAPTER 7

Variance reduction techniques for fixed-source particle transport

CONTENTS

7.1 INTRODUCTION

In Chapter 6, the *analog* Monte Carlo method for a fixed-source problem was introduced, and, in Chapter 5, general variance reduction methodologies for solving integrals or determining integral quantities were discussed. This chapter is devoted to a discussion on the variance reduction techniques developed for fixed-source particle transport problems. in addition to the variance reduction for integral quantities, other techniques including biasing of probability density functions, particle splitting/rouletting, and their combinations are introduced.

The analog Monte Carlo method for particle transport works well except for situations in which the probability of the outcome of interest is low, then significant computation time is needed for achieving a precise and accurate solution. For example, for a simple $1 - D$, homogeneous, purely absorbing shield, with a total cross-section of $\Sigma_t = 2\frac{1}{cm}$ and thickness of $10cm$, the probability of transmission is $\sim 2.1x10^{-9}$. To achieve a precision of 5%, it is necessary to simulate $\sim 1.9x10^{11}$ particle histories. If we consider that each history requires about 10^{-4} seconds, then this *simple* simulation requires ~ 225 days of computation time! This demonstrate that need for more clever techniques referred to as *variance reduction (VR) techniques*.

Over the past 60 years, significant efforts have been devoted to the development of VR techniques. Computer codes, such as [96, 12, 11, 9], MORS [31], MCBEND[23], and TRIPOLI [13] include numerous variance reduction techniques, particularly those techniques based on *pdf* biasing and/or particle splitting and rouletting [24, 30, 25].

Wagner and Haghighat [108, 46] developed the Consistent Adjoint Driven Importance Sampling (CADIS) methodology that is a hybrid technique. It not only combines *pdf* biasing, particle splitting and rouletting, and integration VR techniques, but also it employs the *importance* function obtained from a *deterministic* particle transport code. Further, based on the CADIS methodology, they developed [108, 44] a new version of the MCNP code, referred to *Automated Adjoint Accelerated MCNP* (A³MCNP). A³MCNP enables automation of key tasks including: inputs for a deterministic particle transport code; calculation of importance function; calculation of VR parameters; and, MCNP calculation.

Wagner et al. [110] developed an extension to the CADIS methodology for obtaining global solutions. This extension is referred to as *forward* CADIS (FW-CADIS). Based on CADIS and FW-CADIS,

Wagner and his colleagues developed ADVANTG [109] that provides capabilities similar to A³MCNP.

Other useful references on variance reduction for particle transport include Lux and Koblinger [67], Dunn and Shultis [29], Spanier and Gelbard [94], Okrent et al. [41], and Turner and Larsen [98].

This chapter introduces different variance reduction techniques developed for fixed-source particle transport problems. It gives an overview on variance reduction for particle transport. Four categories of VR techniques are introduced, a select number of techniques in each category are introduced and discussed in detail.

7.2 OVERVIEW OF VARIANCE REDUCTION FOR FIXED-SOURCE PARTICLE TRANSPORT

Variance reduction techniques for particle transport are developed based on changing (*biasing*) of physical quantities or functions for achieving certain objective with reduced variance. To preserve the expected physical outcome, each particle is given a statistical weight (w) that has to be adjusted to compensate for the changed quantity/function. This is accomplished by introducing an equation for the preservation of the *expected outcome* as follows

$$w_{biased} \times (function/quantity)_{biased} = \\ w_{unbiased} \times (function/quantity)_{unbiased} \tag{7.1}$$

The variance reduction techniques used in fixed-source particle transport problems can be grouped into five categories: i) *pdf* biasing; ii) particle splitting and rouletting; iii) weight-window iv) integral biasing; and, v) hybrid methodologies.

A "good" VR technique should result in a precise and accurate solution in a short time. The VR techniques commonly depend on user-defined parameters, which significantly impact the performance and solution accuracy. Hence, any VR technique requires a set of heuristic guidelines and/or theoretical techniques for generation of an *appropriate* set of parameters.

The performance of a MC algorithm with VR is measured based on its FOM. A *good* VR technique is expected to result in significant improvement, i.e. satisfy the following inequality:

$$\frac{FOM_{new}}{FOM_{ref}} \gg 1 \tag{7.2}$$

Note that prior to calculating the FOMs, it is necessary to demonstrate the both algorithms follow the expected behavior of FOM and relative uncertainty as discussed in Chapter 6.

7.3 *PDF* BIASING WITH RUSSIAN ROULETTE

This section elaborates on a select group of variance reduction techniques developed based on the *pdf* biasing technique. Here, to preserve the expected outcome, the *preservation* Equation 7.1 reduces to

$$w_{unbised} \times pdf_{unbiased} = w_{biased} \times pdf_{biased} \tag{7.3}$$

The above formulation indicates that, for example, if the *biased pdf* is increased, the particle weight is decreased proportionally, so that the expected outcome of the *unbiased pdf* is preserved.

The following subsections introduce a few commonly used techniques.

7.3.1 Implicit capture or survival biasing with Russian roulette

In a shielding problem, one of the objectives is to determine the expected detected (survived) particles in the back of the shield. The number of detected particles, however, can be reduced significantly with the increasing absorption interactions, and consequently significant computation time is needed for achieving a statistically reliable result.

In the implicit capture technique, a particle is *forced* to undergo a scattering interaction, by setting the probability of scattering (p_s=1). Using Equation 7.3 for an initial particle weight of w_0, and the unbiased scattering *pdf* ($p_s = \frac{\Sigma_s}{\Sigma_t}$), we obtain the following *preservation* equation

$$w_1 \times 1 = w_0 \times \frac{\Sigma_s}{\Sigma_t} \tag{7.4}$$

Now, we can solve for the biased weight (w_1) as

$$w_1 = w_0 \frac{\Sigma_s}{\Sigma_t} \tag{7.5}$$

Above equation indicates that as p_s is set to 1, i.e, p_s is increased by a factor of $\frac{\Sigma_t}{\Sigma_s}$, the particle weight ($w$) is reduced proportionally by a factor of $\frac{\Sigma_s}{\Sigma_t}$.

Now, if a particle undergoes n interactions, its weight is reduced as follows

$$w_n = w_0 (\frac{\Sigma_s}{\Sigma_t})^n \qquad (7.6)$$

Since Σ_s can be much smaller than Σ_t, Equation 7.6 indicates that the weight of a particle (w_n) may reduce significantly after several interactions. In other words, a low weight particle may have minimal contribution to the particle count and its statistics. In such situations, it is wise to set a lower weight (referred to as *weight cutoff*), at which the particle fate is determined by applying the game of *Russian roulette*.

The implicit capture with Russian roulette is an effective VR technique for deep penetration shielding problems. The Russian roulette technique is described below.

7.3.1.1 Russian roulette technique

After a particle weight has fallen below the weight cutoff, a random number is generated and compared to a parameter, e.g., $\frac{1}{d}$, where d is commonly chosen in a range of $[2, 10]$. Then, the fate of the particle is determined according to:

· if $\eta \leq \frac{1}{d}$, the particle history is continued with the particle weight increased by a factor of d, i.e., $w_r = d \times w$.

· Otherwise, the particle history is terminated.

Note that the Russian Roulette technique can be used with any other VR techniques that encounter very low weight particles.

7.3.2 Path-length biasing

To increase the chance of particle transport to the region of interest (ROI), one may change the free-flight probability such that the particle path-length is stretched toward the ROI (e.g., back of a shield). The FFMC formulation for sampling the path-length (r) (Equation 6.6) indicates that r increases with the decreasing total cross-section. Therefore, in the path-length biasing technique, one may consider replacing the total cross-section with a smaller quantity such as the scattering cross-section. This means that the biased *pdf* for sampling r is expressed by

$$p_{biased}(r) = \Sigma_s e^{-\Sigma_s r} \qquad (7.7)$$

Using the biased *pdf*, the FFMC for sampling path length r reduces to

$$r = -\frac{\ln \eta}{\Sigma_s} \tag{7.8}$$

To preserve the expected outcome for the biased case, using Equation 7.3, we obtain

$$w_1 \times \Sigma_s e^{-\Sigma_s r} = w_0 \times \Sigma_t e^{-\Sigma_t r} \tag{7.9}$$

Then, the biased weight is given by

$$w_1 = w_0 \frac{\Sigma_t}{\Sigma_s} e^{-\Sigma_a r} \tag{7.10}$$

where, $\Sigma_a = \Sigma_t - \Sigma_s$. Here, indeed, the path length (r) is stretched, but it is done toward both sides of the shield. Hence the path length biasing is not effective, and an alternative method is needed to account for direction of ROI.

7.3.3 Exponential transformation biasing

To overcome the limitation of path-length biasing, the exponential transformation has been developed. This technique provides the ability to expand the path-length when a particle is moving toward the ROI, e.g., back of the shield, while contracting the path-length when the it is moving away from the ROI. The biased *pdf* for the exponential transformation is expressed by

$$p_{biased}(r) = (\Sigma_t - c\mu)e^{-(\Sigma_t - \mu)r} \tag{7.11}$$

where c is a user-defined positive parameter, and μ is cosine of the angle between the particle direction and the direction of interest. In choosing the c parameter, one has to maintain that $\Sigma_t - c\mu > 0$ for all particle directions, i.e., $c < \Sigma_t$ (since the maximum positive value for μ is 1). As intended, along the direction of interest ($\mu > 0$), the *pdf* increases, while it decreases for $\mu < 0$. This is further demonstrated if we derive the corresponding FFMC formulation given by

$$r = -\frac{\ln \eta}{\Sigma_t - c\mu} \tag{7.12}$$

In comparison to the path-length biasing formulation, Equation 7.8, Equation 7.12 yields a larger path-length only in the direction of interest, while it shortens the path-length in the opposite direction. To preserve the expected outcome, Equation 7.3 for the exponential transformation technique reduces to

$$w_1 \times (\Sigma_t - c\mu)e^{-(\Sigma_t - c\mu)r} = w_0 \times \Sigma_t e^{-\Sigma_t r} \qquad (7.13)$$

Therefore, the formulation for the biased weight is expressed by

$$w_1 = w_0 \frac{\Sigma_t}{\Sigma_t - c\mu}e^{-c\mu r} \qquad (7.14)$$

Now, if a particle goes through n scattering interactions followed by n path-length sampling, the biased weight reduces to

$$w_n = w_0 e^{-c(x-x_0)} \prod_{i=1}^{n} \frac{\Sigma_{t,i}}{\Sigma_{t,i} - c\mu_i}, \qquad (7.15)$$

where $x - x_0 = \sum_{i=1}^{n} \mu_i r_i$, and x_0 and x refer to the initial and final positions of the particle on the axis of interest (e.g., $x - axis$), respectively. Note that if $(\Sigma_{t,i} - c\mu_i)$ becomes very small, then the biased weight may become very large, resulting in a large count, and potentially a biased solution. To remedy this situation, this method should be combined with a splitting technique such as the weight-window technique that will be discussed later.

Finally, it is important to note that this technique can be easily combined with the *implicit capture* technique, as both are effective for a shielding problem.

7.3.4 Forced collision biasing

This technique forces particle collisions in a small volume that otherwise can be missed using the analog particle transport. The technique is described as follows:

1. A particle of weight (w) enters the volume of interest.

2. The particle is split into two particles (*uncollided* and *collided*) of smaller weights: the *uncollided* particle passes through the volume without any collisions; the *collided* particle is forced to have a collision in the volume.

3. The weight of the *unollided* particle is set to

$$w_{uncollided} = we^{-\Sigma_t u} \qquad (7.16)$$

and the weight of the *collided* particle is set to

$$w_{collided} = w(1 - e^{-\Sigma_t u}) \tag{7.17}$$

Here, u refers the path-length that spans the volume along the particle direction.

4. The *pdf* (referred to as $g(r)$) for sampling the path-length of the *collided* particle within the volume of interest is obtained by setting the preservation equation (Equation 7.3) as follows

$$w_{collided} \times g(r) = w \times \Sigma_t e^{-\Sigma_t r} \tag{7.18}$$

Hence, the *pdf* for the *collided* particle reduces to

$$g(r) = \frac{\Sigma_t e^{-\Sigma_t r}}{1 - e^{-\Sigma_t u}} \tag{7.19}$$

5. Path-length (r) within the volume for the collided particle is sampling by using the following FFMC derive based on the $g(r)$ probability density function.

$$r = -\frac{\ln[1 - \eta(1 - e^{-\Sigma_t u})]}{\Sigma_t} \tag{7.20}$$

7.4   PARTICLE SPLITTING WITH RUSSIAN ROULETTE

The concept of particle splitting stems from the fact that different particles (themselves or through their progenies) have different levels of contribution (or *importance*) to an objective. In other words, in a simulation, there are particles of various degrees of *importance*. Therefore, it is desirable to increase the survivability of the *important* particles, and simultaneously judiciously eliminate *less important* particles. It is obvious that a splitting technique can be effective only if the importance of particles are selected *appropriately*. Using erroneous *importance* may lead to biased results in a rather short time, or waste of computation time by splitting and transporting *unimportant* (or *low weight*) particles. In this section, we will discuss particle splitting techniques based on particle attributes including position, energy, direction, and weight.

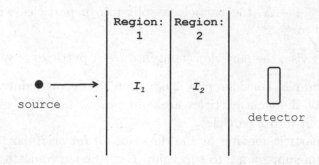

Figure 7.1 Schematic of a 1-D shield for demonstration of the geometric splitting

7.4.1 Geometric splitting

In this technique, the simulation region is partitioned into a number of subregions, and each subregion is given an importance (I) such that the importance increases as one approaches the region of interest. As a particle moves from a subregion of *low importance* to a subregion of *high importance*, it is split into a number of particles with lower importance; thereby, improving the chance of sampling *good* particles (which are moving toward the ROI). Conversely, if a particle travels from a *high importance* subregion to a *low importance* subregion, a game of *Russian roulette* is performed.

This technique can be explained more clearly via a 1-D shield that is partitioned into two subregions with importances of I_1 and I_2, respectively (as shown in Figure 7.1). If our objective is to estimate the probability of transmission through the shield, then it is necessary to set $I_1 < I_2$.

For a particle of weight w moving in the direction of higher importance, i.e., from subregion 1 to sub region 2, the particle is split into a number of identical particles of lower weights depending on the importance ratio of subregions, i.e., $\frac{I_2}{I_1}$ as follows:

1. If $n = \frac{I_2}{I_1}$ is an integer ≥ 2, the particle is split into n identical particles, each with a weight of $\left(\frac{w}{n}\right)$.

2. If $r = \frac{I_2}{I_1}$ is a real number ≥ 2, a random number η is generated and compared with

$$\Delta = \frac{I_2}{I_1} - INT\left[\frac{I_2}{I_1}\right] = r - n \qquad (7.21)$$

a. if $\eta \leq 1 - \Delta$, the particle is split into n particles with weight equal to $\frac{w}{n}$.

b. Otherwise, the particle is split into $n+1$ particles of weight $\frac{w}{n+1}$.

For example, consider $r = 2.63$, then above procedure indicates that 63% of the time particles are split into 3, and 37% of the time they are split into 2 particles.

For a particle moving in the direction of *lower importance*, i.e., moving from subregion 2 to subregion 1, a Russian roulette procedure is applied as follows:

1. A random number η is generated.

2. If $\eta \leq \frac{1}{r}$, the particle kept and its weight is increased to $w \times r$.

3. Otherwise, the particle is killed.

To assure that the geometric splitting technique is effective, one has devote significant care in selecting appropriate importance for each subregion. A few heuristic recommendations [96] are listed below:

1. Keep the ratio of region importances close to each other (within a factor of 2–4).

2. Consider region sizes of 2 mfp.

3. Maintain the same number of particles for all regions.

4. Do not split particles in a void.

The major difficulty in using the geometric splitting technique is in assigning the *proper importance* to subregions. This issue can be significantly remedied, if one knows the neutron importance (or adjoint) function. Wagner and Haghighat [108] have developed methodologies for automatic generation of the space- and energy-dependent importance distributions. The importance function is discussed in Section 7.6.1.

It also is worth noting that, in highly angular dependent situations, the effectiveness of geometric splitting may diminish because its importance is not angular-dependent. The use of an angular-dependent importance function may remedy this issue. Finally, a zero importance can be assigned to a region; this is quite useful in representing non-reenterant boundary conditions and/or performing model simplifications/changes.

7.4.2   Energy splitting

In some simulations, certain energy ranges are more important than others, thus, we may split particles when their energies lie within the important energy ranges. Again Russian roulette is used for particles that may not contribute significantly to the problem objective. As an example, if there are two isotopes that compete in neutron capture (at different energies) and one of them is of interest, we split the particles (with "correct" energy) so that we can increase the chance of interaction with the isotope of interest.

7.4.3   Angular splitting

If a particular range of solid angle is important in a simulation, we may split particles when they move into that solid angle. Russian roulette is used for particles that move out of the solid angle.

7.5   WEIGHT-WINDOW TECHNIQUE

This technique combines space, energy, and angular splitting techniques with the Russian rouletting, but commonly, only space- and energy-splitting are considered. The weight-window provides a unique technique for controlling both high and low particle weights; high weight particle may result in biasing the solution, while low weight particle may lead to significant inefficiency, as low weight particle may have minimal impact on the expected objective.

Partitioned into space-energy cells, and a weight window with lower and upper weights, $w_\ell(\overrightarrow{r}, E)$ and $w_u(\overrightarrow{r}, E)$ is assigned to each cell. If a particle at position $\overrightarrow{r}$ with energy E has a weight in the acceptable range of $[w_\ell, w_u]$, its history is continued; otherwise, it is treated via one of the following procedure:

a. If the particle weight is less than w_ℓ, Russian roulette is performed and the particle is either terminated or its weight is increased to a value (w_s) within the acceptable range.

b. If the particle weight is greater than w_u, the particle is split into new particles of weight within the acceptable range.

Generally, w_u and w_s are related to the lower weight (w_ℓ). For example, the MCNP code [96] recommends the following relations:

$$w_u = 5w_\ell \qquad (7.22)$$

and

$$w_s = 2.5w_\ell \qquad (7.23)$$

Commonly, *proper* selection of an appropriate weight distribution for a complex problem is difficult and requires a significant amount of an analyst's time. The MCNP code [96] offers an iterative procedure for the generation of weights, referred to as the *weight−window generator* [8, 10, 49]. However, Haghighat strongly recommends the use of the *importance (adjoint) function* methodology for determination of the energy and space-dependent weight distributions. This is discussed in detail in the following section.

7.6   INTEGRAL BIASING

In a shielding problem, one of the key objectives is determination of the radiation dose or detector response at the surface of a shield. Theoretically, the detector response can be calculated by

$$R = < \Sigma_d \psi > \qquad (7.24)$$

where, $<>$ refers to the Dirac sign that refers to integration over all independent variables $(\overrightarrow{r}, E, \hat{\Omega})$, R refers to detector response, Σ_d is detector cross-section, and ψ refers to particle angular flux that can be calculated using the linear Boltzmann equation (6.1). Another approach to determine the dose or response, is to use the *importance (adjoint) function* methodology discussed in the following subsection.

7.6.1   Importance (adjoint) function methodology

The linear Boltzmann Equation 6.1 for a fixed-source problem can be written in an operator form as follows

$$H\psi(p) = q(p), \; in \; V \qquad (7.25)$$

where, p refers to $(\overrightarrow{r}, E, \hat{\Omega})$, V refers to the problem volume, ψ is angular flux, q is the fixed source distribution, and H is the transport operator given by

$$H = \hat{\Omega} \cdot \nabla + \Sigma_t - \int_0^\infty dE' \int_{4\pi} d\Omega' \Sigma_s(\overrightarrow{r}, E \to E', \hat{\Omega} \to \hat{\Omega}') \qquad (7.26)$$

Considering a vacuum (non-reentrant) boundary condition, the angular flux for incoming directions is expressed as

$$\psi = 0, \;\; for \;\; \hat{n} \cdot \hat{\Omega} < 0 \;\; and \;\; \vec{r} \epsilon \; \Gamma \tag{7.27}$$

The corresponding particle importance equation [4] is given by

$$H^* \psi^*(p) = q^*(p), \;\; in \;\; V \tag{7.28}$$

where, ψ^* is the particle importance function, q^* is importance source, and H^* is the importance operator given by

$$H^* = -\hat{\Omega} \cdot \nabla + \Sigma_t - \int_0^\infty dE' \int_{4\pi} d\Omega' \Sigma_s(\vec{r}, E' \to E, \hat{\Omega}' \to \hat{\Omega}) \tag{7.29}$$

and the importance function for the vacuum boundary condition is expressed as

$$\psi^* = 0, \;\; for \;\; \hat{n} \cdot \hat{\Omega} > 0 \;\; and \;\; \vec{r} \epsilon \; \Gamma \tag{7.30}$$

Now, we form the *commutation relation* between Equations 7.25 and 7.28, we obtain

$$< \psi^* H \psi > - < \psi H^* \psi^* > = < \psi^* q > - < \psi q^* > \tag{7.31}$$

Considering a vacuum boundary condition, one can show the left-hand side of above equation is equal zero, hence

$$< \psi q^* > = < \psi^* q > \tag{7.32}$$

Now, we consider that the importance source is given by

$$q^* = \Sigma_d \tag{7.33}$$

Hence, the detector response (R) equation (Equation 7.24) can be written in terms of the importance function as

$$R = < \psi^* q > \tag{7.34}$$

where, ψ^* is obtained from Equation 7.28 for a given detector, and q is a given source distribution.

7.6.2   Source biasing based on the importance sampling

In a Monte Carlo shielding simulation, the normalized source distribution (source *pdf*) is sampled, the sampled particles are transported, then a fraction of survived particles are detected depending on the detector cross-section. The detected particles effectively represent the detector response (R)(Equation 7.34). This means that if ψ^* function was known, then the integral is solved by sampling the normalized source as the *pdf*. To reduce the variance of the sampled integral, the importance sampling can be very effective. From importance sampling, the most effective *pdf* is expressed by

$$\hat{q}(p) = \frac{\psi^*(p)q(p)}{R} \tag{7.35}$$

Hence, from this observation we conclude that for a Monte Carlo shielding simulation, a more effective *pdf* is a biased source using an approximate importance function as follows

$$q_{biased}(p) = \frac{\psi_a^*(p)q(p)}{R_a} \tag{7.36}$$

where, ψ_a^* is the approximate importance function that is obtained by solving an approximate form of Equation 7.28, and the corresponding approximate response (R_a) is obtained by solving Equation 7.34. Now, to reserve the expected source, we form the corresponding *preservation* equation as follows

$$w_{biased}(p) \times q_{biased}(p) = w(p) \times q(p) \tag{7.37}$$

then, the biased source particle weight is obtained by

$$w_{biased}(p) = \frac{q(p)}{\frac{q(p)\psi_a^*(p)}{R_a}} = \frac{R_a}{\psi_a^*(p)} \tag{7.38}$$

7.7   HYBRID METHODOLOGIES

Over the past 20 years, there has been significant effort on the development of hybrid methodologies that employ deterministic technique for determination of VR reduction parameters and/or biased functions. Here, we will introduce the CADIS (consistent adjoint driven importance sampling) methodology [106, 108, 46], and its extension the *forward*-CADIS (FW-CADIS) [109, 78].

7.7.1 CADIS methodology

The CADIS methodology combines source biasing and space-energy splitting/rouletting within the weight-window technique. The biased source and lower weight (w_ℓ) are determined in a consistent manner by using an approximate importance function obtained from a *deterministic* discrete-ordinates particle transport calculation. Further, to make the method practical, only space and energy dependencies are considered. The biased source is given by

$$\hat{q}(\overrightarrow{r}, E) = \frac{\phi_a^*(\overrightarrow{r}, E)q(\overrightarrow{r}, E)}{R_a} \tag{7.39}$$

And the corresponding biased weight for source particles is expressed by

$$\hat{w}(\overrightarrow{r}, E) = \frac{R}{\phi_a^*(\overrightarrow{r}, E)} \tag{7.40}$$

Using Equations 7.22 and 7.23, we obtain the lower weight of the weight window as follows

$$w_s(\overrightarrow{r}, E) = \frac{R_a}{\phi_a^*(\overrightarrow{r}, E)} \tag{7.41}$$

then, since w_s is given by

$$w_s = w_\ell \left(\frac{1+c}{2}\right) \tag{7.42}$$

the lower weight is given by

$$w_\ell(\overrightarrow{r}, E) = \left(\frac{2}{1+c}\right)\frac{R_a}{\phi_a^*(\overrightarrow{r}, E)} \tag{7.43}$$

Further, during particle transport between $(\overrightarrow{r}, E)$ and $(\overrightarrow{r}', E'))$ phase spaces, particles are split if the importance ratio of $\frac{\phi^*(\overrightarrow{r}, E)}{\phi^*(\overrightarrow{r}', E')})$ greater than 1, otherwise they are rouletted. Consequently, the particle weight is adjusted according to

$$w(\overrightarrow{r}, E) = w(\overrightarrow{r}', E')\frac{\phi^*(\overrightarrow{r}', E')}{\phi^*(\overrightarrow{r}, E)} \tag{7.44}$$

It is worth noting that, based on the CADIS methodology, a modified version of MCNP code called A³MCNP [108, 46] has been developed, which automatically determines the importance function distributions, the biased source, and the weight-window lower weights. Wagner [107] has developed a similar algorithm referred to as ADVANTG. Note that the Monaco Monte Carlo code within the SCALE package uses the CADIS formulation through a processing code referred to as MAVRIC [81].

7.7.1.1 FW-CADIS technique

The CADIS methodology is very effective for the determination of localized objectives; however, it is not effective for determination of distributed response, e.g., radiation dose distribution throughout a power plant. Wagner, Peplow, and Mosher [109] have developed a novel extension to the CADIS, referred to as FW-CADIS, in which the importance source is weighted inversely by the spatial-dependent forward flux, e.g.,

$$q^*(\overrightarrow{r}, E) = \frac{\Sigma_d(\overrightarrow{r}, E)}{\int_0^\infty dE' \Sigma_d(\overrightarrow{r}, E') \phi(\overrightarrow{r}, E')} \tag{7.45}$$

where $phi(\overrightarrow{r}, E)$ the space- and energy-dependent scalar flux. This formulation results in an increase of the adjoint source in regions with low flux values, and vice versa. In other words, the division by the forward flux increases the importance of regions away from the source. FW-CADIS has proved very effective when the volume is large, e.g., a nuclear reactor, and it is necessary to determine a certain quantity, e.g., dose, throughout the model. Note that, if necessary, angular dependency can be included. However, more detail can require large amount of memory, and, therefore, may prevent the use of the method or reduce the overall benefit. For further details on the methodology and its application, the reader should consult [110, 109]. Note that FW-CADIS is included in the MAVRIC processing routing of the SCALE/ Monaco code system.

7.8 REMARKS

This chapter discusses the variance reduction techniques used in the fixed-source Monte Carlo particle transport simulations. It groups the variance reduction techniques into five groups including: i) *pdf* biasing; ii) particle splitting biasing; iii) weight-window; iv) integral biasing; and, v) hybrid methodologies. Commonly, for shielding problems, the implicit capture technique is used as a default VR technique, because it eliminates one of two major mechanisms for particle loss. The CADIS hybrid methodology and its extension FW-CADIS, combine several biasing techniques including implicit capture, source biasing, weight-window techniques, and utilize the deterministic methods for the determination of VR parameters. The A^3MCNP and ADVANTG code systems can significantly benefit users as no only effectively and consistently combine different techniques, but also provide physics-based automation algorithms for generation of VR parameters. This latter aspect of the code significantly reduce the chance of solution biasing and/or long computation times.

PROBLEMS

1. Modify the program developed in Chapter 6 Problem 4 to evaluate precision and FOM associated with the probabilities of transmission, reflection, and absorption of a source through a 1-D slab. Consider the following two cases:

   a. $\Sigma_t d = 10$, $\dfrac{\Sigma_s}{\Sigma_t} = 0.2$, and $\theta = 0.0, 30.0, 60.0$

   b. $\Sigma_t d = 10$, $\dfrac{\Sigma_s}{\Sigma_t} = 0.8$, and $\theta = 0.0, 30.0, 60.0$

   where d is the shield thickness and θ is the source direction. To stop your simulation, set the maximum relative uncertainty (R) equal to 10% and the maximum number of experiments to 10 million. Use Section 4.7 to determine the relative uncertainty and use the maximum relative uncertainty to determine the FOM. Use Table 7.1 to present your results for each case.

Table 7.1 A sample table for problem 1

θ	Number of histories	Transmission (%)	R_{trans}	Reflection (%)	R_{refl}	Absorption (%)	R_{abs}	FOM
0.0								
30.0								
60.0								

2. Modify the program in Problem 1 for inclusion of implicit capture. Perform similar analysis as Problem 1, and compare the results to those from Problem 1.

3. Modify the program in Problem 1 for inclusion of geometric splitting. Perform similar analysis as Problem 1, and compare the results to those from Problems 1 and 2.

4. Consider a 1-D slab of thickness 10 cm, located in a vacuum region, and a planar fixed-source is placed on its left boundary. The fixed-source emits particles only normal to the surface, and the slab contains a homogeneous material with total cross-section is 1.0 cm^{-1} and absorption cross-section of 0.5 cm^{-1}.

   a. The one-speed, 1-D importance diffusion equation is expressed by:

   $$-D\frac{d^2\Phi^*}{dx^2} + \Sigma_a\Phi^* = S^*, \quad 0 < x < L, \quad \text{where } D = \frac{1}{3\Sigma_t}$$

   Use the 1-D importance diffusion equation to determine the importance function distribution for particles to leave the right side of the slab. Since the goal is to have particles leave the right side of the system, we have $S^* = 0$, but we set the adjoint boundary current to be equal to 1 for particles entering from the right, i.e.:

   $$D\frac{d\Phi^*}{dx}\bigg|_{x=L} = 1$$

   On the left boundary, use a vacuum boundary condition, i.e.:

   $$\Phi^*\bigg|_{x=-2D} = 0$$

   b. Determine average particle importance per mean free path section, i.e., 1.0 cm based on part (a) results.

   c. Use the program you developed in Problem 3 with the importance distribution given in part (b). Examine the code's performance for determination of transmission probability while achieving a relative error of 5%. Determine FOM for this simulation.

   d. Repeat parts (b) and (c) for absorption cross sections of 0.7 cm^{-1} and 0.9 cm^{-1}, considering that the total cross-section remains the same.

5. Consider the slab in Problem 4 contains three regions; an absorbing region is surrounded by two scattering regions. The absorbing region is 4 cm-thick, and each scattering region has a thickness of 3 cm. The scattering ratio is 0.90 cm^{-1} in the scattering regions, and is 0.10 cm^{-1} in the absorbing region.

a. Just like Problem 4, determine the adjoint (importance) function.

b. Determine average particle importance per mean free path based on part (a) results.

c. Use the program you developed in Problem 4 with the importance distribution given in part (b), and examine the performance of the code for determination of transmission probability for achieving a relative uncertainty of 5%. Determine FOM for this simulation.

6. Based on the FOM, examine the performance of your code in Problem 3 if you combine the implicit capture and geometric splitting techniques.

7. Consider a slab with a planar isotropic neutron source placed in its center. If the slab is placed in a vacuum, determine the probability of leakage from the slab. The slab contains a material of total cross-section 1.0 cm^{-1} and absorption cross-section of 0.5 cm^{-1}. Considering that the slab thickness is 20 cm:

a. Use 1-D diffusion equation to determine an importance function corresponding to a planar detector placed at each boundary.

b. Determine averaged particle importance per mean free path based on the right- and left-boundary detectors.

c. Use the program you developed in Problem 3 with the two importance distributions obtained in part (b), and examine the performance of the code for the determination of the transmission probability for achieving a relative uncertainty of 5%. Determine FOM for this simulation.

8. Repeat Problem 7 if the absorption cross-section in the first 10 cm is 0.7 cm^{-1}, and 0.3 cm^{-1} in the second 10 cm.

9. Consider a detector is placed in a material region that is surrounded by vacuum. If there is no source inside the region, but there is an isotropic boundary source, derive a formulation for determination of detector response in terms of the adjoint function. Follow the derivations given in Section 7.6.1. Write a formulation for the biased source and the corresponding weight.

Scoring/Tallying

CONTENTS

8.1 INTRODUCTION

In a particle transport problem, in addition to estimating the number of particles transmitted, reflected, or absorbed, one is interested in estimating different physical quantities, such as flux, current, and reaction rates of particles. For design and analysis of a steady-state system, it is necessary to determine these quantities in a phase space $(d^3 r dE d\Omega)$ and, in time-dependent situations, variations of the phase space information in a time differential (dt) must be estimated. Therefore, in general, one has to partition the Monte Carlo model into space (ΔV), energy(ΔE), angle $(\Delta \Omega)$, and/or time (Δt) intervals, and tally the particles moving into these intervals. The MCNP (Monte Carlo N-Particle) code manual [96], Foderaro [35], and Lewis and Miller [65] provide excellent discussions on tallying options, formulations, and related uncertainties.

In this chapter, we will introduce different tally estimators and derive steady-state formulations for different quantities using different estimators. Time-dependent tallying will be discussed and the change in steady-state formulations will be demonstrated. Finally, we derive formulations to determine uncertainties associated with the estimated quantities and address uncertainties of random variables which depend on several other random variables, i.e., error propagation.

8.2 MAJOR PHYSICAL QUANTITIES IN PARTICLE TRANSPORT

In a particle transport simulation, there are three major physical quantities, including *flux, current, and reaction rate*. All of these are related to the expected number of particles in the phase space at time t, defined as:

$$n(\overrightarrow{r}, E, \Omega, t)d^3rdEd\Omega \equiv Expected \ number \ of \ particles$$
$$in \ d^3r \ about \ vector \ position \ \overrightarrow{r},$$
$$moving \ in \ d\Omega \ about \ \hat{\Omega}, \ with \ energy \tag{8.1}$$
$$within \ dE \ about \ E, \ at \ time \ t,$$

where the unit of $n(\overrightarrow{r}, E, \Omega, t)$ is $\left(\frac{1}{cm^3-eV-steradian}\right)$.

We define angular flux as the product of angular neutron density n and particle speed (v) expressed by

$$\psi(\overrightarrow{r}, E, \hat{\Omega}, t) = v(E)n(\overrightarrow{r}, E, \hat{\Omega}, t), \tag{8.2}$$

where the unit of the angular flux is $\left(\frac{1}{cm^2-eV-steradian-sec}\right)$. The angular current density is defined by

$$\overrightarrow{j}(\overrightarrow{r}, E, \hat{\Omega}, t) = \hat{\Omega}\psi(\overrightarrow{r}, E, \hat{\Omega}, t) \tag{8.3}$$

However, in most practical simulations, one is interested in integral quantities, such as scalar flux, partial currents, and reaction rates, that are independent of the angular variable. This is because it is very expensive and impractical to obtain angular-dependent quantities, either experimentally or computationally; moreover, for most situations, angular-dependent quantities have limited use.

The scalar flux is defined as the integral of angular fluxes over all directions, defined by

$$\phi(\overrightarrow{r}, E, t) = \int_{4\pi} d\Omega \psi(\overrightarrow{r}, E, \hat{\Omega}, t). \tag{8.4}$$

where the unit of the scalar flux is $\left(\frac{1}{cm^2 - eV - sec}\right)$.

The positive and negative partial currents are defined as

$$j_{\pm}(\overrightarrow{r}, E, t) = \int_{2\pi\pm} d\Omega \hat{n} \cdot \hat{\Omega} \psi(\overrightarrow{r}, E, \hat{\Omega}, t). \tag{8.5}$$

where $\hat{n}$ is the unit vector along the direction of interest, commonly the normal vector to a surface. The partial currents are calculated for estimating the number of particles moving in the positive or negative sense of a surface.

The reaction rate per unit volume of a particular type c is determined using

$$R_c(\overrightarrow{r}, t) = \int_0^{\infty} dE \Sigma_c(\overrightarrow{r}, E)\phi(\overrightarrow{r}, E, t) \tag{8.6}$$

where $\Sigma_c(\overrightarrow{r}, E)$ refers to the type c cross-section at position vector $\overrightarrow{r}$ and particle energy of E, and the unit of R is ($\frac{1}{cm^3 - sec}$).

8.3 TALLYING IN A STEADY-STATE SYSTEM

In a steady-state system it is necessary to determine the number of particles in the phase space. For example, one may partition the Monte Carlo model into I spatial volumes (ΔV_i), J energy intervals (ΔE_j), and K angular bins (Ω_k) to perform particle tallying (counting).

Four different techniques are commonly used for tallying or scoring, these include:

1. Collision estimator

2. Path-length estimator

3. Surface-crossing estimator

4. Analytical estimator

The following sections will describe each technique and discuss their use and limitations.

8.3.1 Collision estimator

First, we partition (discretize) spatial, energy, and angular domains into I, J, and K intervals (grids), respectively. Then, we score or count any particle of weight w moving along direction $\hat{\Omega}$ within $\Delta\Omega_k$, with energy E within ΔE_j, and has a collision in ΔV_i.

A collision counter array $C(i, j, k)$ is increased by the particle weight as

$$C(i, j, k) = C(i, j, k) + w \tag{8.7}$$

Note that w is equal to 1 if no variance reduction is considered.

After H particle histories (sources), the normalized collision density is given by

$$g(\overrightarrow{r_i}, E_j, \hat{\Omega}_k) = \frac{C(i, j, k)}{H \Delta V_i \Delta E_j \Delta\Omega_k} \left(\frac{\frac{\#collisions}{cm^3 - eV - Steradian - sec}}{\frac{\#source-particle}{sec}} \right) \tag{8.8}$$

and the angular flux is given by

$$\psi(\overrightarrow{r_i}, E_j, \hat{\Omega}_k) = \frac{g(\overrightarrow{r_i}, E_j, \hat{\Omega}_k)}{\Sigma_t(E_j)} = \frac{C(i, j, k)}{H \Delta V_i \Delta E_j \Delta\Omega_k \Sigma_t(E_j)}$$
$$\left(\frac{\frac{\#}{cm^2 - eV - Steradian - sec}}{\frac{\#source-particle}{sec}} \right) \tag{8.9}$$

Note that there is a fundamental difficulty with Equation 8.9; it is divided by a cross-section value at E_j rather than the particle energy E, which is not known, as the division after accumulation of counts. This is important because cross-section may change significantly within ΔE_j, therefore the calculated angular can be erroneous. To overcome this difficulty, we define a new counter given by

$$FC(i, j, k) = FC(i, j, k) + \frac{w}{\Sigma_t(E)} \tag{8.10}$$

Therefore, the angular flux is expressed by

$$\psi(\overrightarrow{r_i}, E_j, \hat{\Omega}_k) = \frac{FC(i, j, k)}{H \Delta V_i \Delta E_j \Delta\Omega_k} \tag{8.11}$$

and the scalar flux is given by

$$\phi(\overrightarrow{r_i}, E_j) = \sum_{k=1}^{K} \psi(\overrightarrow{r_i}, E_j, \hat{\Omega}_k) \Delta\Omega_k = \frac{\sum_{k=1}^{K} FC(i, j, k)}{H \Delta V_i \Delta E_j} \tag{8.12}$$

To determine the reaction rate, we introduce another counter given by

$$CC(i,j) = CC(i,j) + w\frac{\Sigma_c}{\Sigma_t} \tag{8.13}$$

and, therefore, the reaction rate density is expressed by

$$R(\overrightarrow{r_i}) = \frac{\sum_{j=1}^{J} CC(i,j)}{H\Delta V_i} \tag{8.14}$$

It is important to note that the collision estimator is not efficient if the probability of interaction (i.e., Σ_t) is low, i.e., optically thin media.

8.3.2 Path-length estimator

The path-length estimator is very effective in optically thin regions, where the collision estimator is not efficient or practical. This estimator is derived based on the fact that the particle flux can be defined as the total path-length of particles traveling aling different directions per unit volume.

The path-length estimator is defined as follows: any particle of weight w moving in direction $\hat{\Omega}$ within $\Delta\Omega_k$, with energy E within ΔE_j, which traces a path-length(p) within volume ΔV_i is counted. Accordingly, a counter for this estimator can be defined as

$$p(i,j,k) = p(i,j,k) + w \times p \tag{8.15}$$

Note that different particle path-lengths may have different starting and ending points relative to a volume, as shown in Figure 8.1, that is,

1. Path-length starts in the volume and ends outside the volume.

2. Path-length starts and ends in the volume.

3. Path-length starts outside the volume and ends inside the volume.

4. Path-length crosses the volume.

Now, we may use the $p(i,j,k)$ counter to estimate the angular flux given by

$$\psi(\overrightarrow{r_i}, E_j, \hat{\Omega}_k) = \frac{p(i,j,k)}{H\Delta V_i \Delta E_j \Delta\Omega_k} \left(\frac{\frac{cm}{cm^3 - eV - steradian - sec}}{\frac{\#}{sec}} \right), \tag{8.16}$$

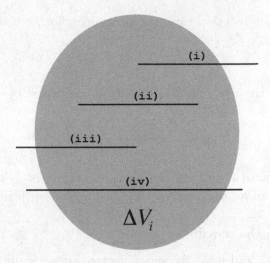

Figure 8.1 Possible particle traces for the path-length estimator

and the scalar flux is given by

$$\phi(\vec{r_i}, E_j) = \sum_{k=1}^{K} \frac{p(i,j,k)}{H\Delta V_i \Delta E_j \Delta \Omega_k} \left(\frac{\frac{cm}{cm^3 - eV - s}}{\frac{\#}{sec}} \right) \qquad (8.17)$$

To obtain the reaction rate per unit volume, we define another counter given by

$$CP(i,j) = CP(i,j) + w \times p \times \Sigma_c(E) \qquad (8.18)$$

where Σ_c is the cross-section for interaction type c. Then, the formulation for the reaction rate is expressed by

$$R_c(\vec{r}_i) \frac{\sum_{j=1}^{J} CP(i,j)}{H\Delta V_i} \left(\frac{\frac{\#}{cm^3 - sec}}{\frac{\#}{sec}} \right) \qquad (8.19)$$

8.3.3 Surface-crossing estimator

Both the collision and path-length estimators count particles over a cell volume. In order to estimate surface-wise information, it necessary to reduce the thickness of the volume cell. This would result in the loss of precision, because fewer particles can travel through the volume of interest. In order to overcome this difficulty, we devise another technique called the surface-crossing estimator to estimate the current and, possibly the scalar flux.

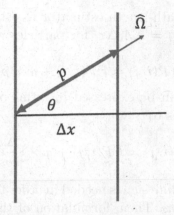

Figure 8.2   Estimation of scalar flux using the surface-crossing estimator

8.3.3.1   Estimation of partial and net currents

To estimate energy-dependent current densities, we introduce a counter to account for any particle of weight w, with direction Ω and energy E within ΔE_j, that crosses the surface area ΔA_i. To distinguish between the particles moving from left to right versus those moving from right to left, we define a counter as

$$SC(i,j,kk) = SC(i,j,kk) + w \times |\hat{n} \cdot \hat{\Omega}| \qquad (8.20)$$

where kk is either 1 or 2, referring to *positive* or *negative* senses of the surface, respectively. Hence, the *positive* and *negative* partial currents are given by

$$J_{\pm}(\vec{r_i}, E_j) = \frac{SC(i,j,kk)}{H \Delta A_i \Delta E_j} \left(\frac{\frac{\#}{cm^2 - eV - sec}}{\frac{\#}{sec}} \right), \quad for \ \ kk = 1 \ \ or \ \ 2 \quad (8.21)$$

The net current is expressed by

$$J_{net}(\vec{r_i}, E_j) = J_{+}(\vec{r_i}, E_j) - J_{-}(\vec{r_i}, E_j) \qquad (8.22)$$

8.3.3.2   Estimation of flux on a surface

The surface-crossing estimator also can be used (with *caution*) to estimate the scalar flux, which is a volumetric quantity. To do so, consider a thin foil of thickness Δx and area ΔA as shown in Figure 8.2.

We may use the path-length estimator to estimate the scaler flux within the volume $\Delta V_i = \Delta A_i \Delta x_i$ for particles with energy E within ΔE_j as

$$FP(i,j) = FP(i,j) + w \times p \qquad (8.23)$$

Using Figure 8.2, p can be expressed in terms of Δx_i, and therefore above equation reduces to

$$FP(i,j) = FP(i,j) + w \times \frac{\Delta x_i}{|cos\theta|} \qquad (8.24)$$

Note that the *absolute* sign is needed in order to include all particles moving in all directions. Then, formulation of the scalar flux reduces to

$$\phi(\overrightarrow{r_i}, E_j) = \frac{FP(i,j)}{H \Delta V_i \Delta E_j}$$

or

$$\phi(\overrightarrow{r_i}, E_j) = \frac{FP(i,j)}{H \Delta A_i \Delta x_i \Delta E_j}$$

or $\qquad (8.25)$

$$\phi(\overrightarrow{r_i}, E_j) = \frac{\frac{FP(i,j)}{\Delta x_i}}{H \Delta A_i \Delta E_j}$$

or

$$\phi(\overrightarrow{r_i}, E_j) = \frac{FS(i,j)}{H \Delta A_i \Delta E_j}$$

where, we have defined a new counter for determination of the scalar flux using the surface-crossing estimator given by

$$FS(i,j) = FS(i,j) + w \times \frac{1}{|cos\theta|} \qquad (8.26)$$

The difficulty for using the Equation 8.26 for estimation of the flux is that the counter approaches infinity as θ approaches 90°. Hence, commonly an exclusion angle of a few degrees is considered, e.g., the MCNP code uses 3°.

8.3.4 Analytical estimator

The aforementioned three tallying estimators are based on a segment of volume or area, henceforth they cannot provide an efficient capability

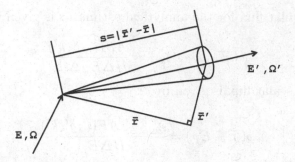

Figure 8.3 Schematic of particle transport for the analytical estimator

for tallying over small or point-like regions. The analytical estimator is an attempt to overcome this shortcoming.

This estimator can be described as follows: as a particle is born or undergone a scattering interaction, the probability that it will be detected at a point of interest is evaluated analytically. Figure 8.3 depicts the particle position and its relation to the point of interest.

The probability that a particle will be detected at the point of interest is composed of two independent probabilities:

1. $p(E \to E', \mu_0)d\Omega'dE' \equiv$ Probability that a particle of energy E, moving in direction $\hat{\Omega}$, will be scattered into $\Delta\Omega'$ about Ω' with energy E' within $\Delta E'$.

2. $e^{-\tau(\vec{r}, \vec{r}', E')} \equiv$ probability that a particle of energy E', moving in direction $\hat{\Omega}'$ will survive a distance $s = |\vec{r}' - \vec{r}|$, where τ is expressed by $\int_0^s ds' \Sigma_t(s', E')$.

Hence, the combined probability that a particle of energy E moving along direction $\hat{\Omega}$ will be detected at vector position $\vec{r}'$ is given by

$$p(E \to E', \mu_0)e^{-\tau(\vec{r}, \vec{r}', E')}d\Omega'dE' \qquad (8.27)$$

Now, if the above equation is divide by dA_r, and $d\Omega'$ is substituted by $\frac{dA}{r^2}$, we obtain a formulation for particle count per unit area as follows

$$F = p(E \to E', \mu_0)e^{-\tau(\vec{r}, \vec{r}', E')}\frac{1}{s^2}dE' \qquad (8.28)$$

Hence, the counter for the analytical estimator is expressed by

$$AF(i', j', k') = AF(i', j', k') + w \times F \qquad (8.29)$$

The angular flux for the analytical estimator is given by

$$\psi(\overrightarrow{r}_{i'}, E_{j'}, \Omega_{k'}) = \frac{AF(i', j', k')}{H \Delta E_{j'} \Delta \Omega_{k'}}, \tag{8.30}$$

and the scalar flux is given by

$$\phi(\overrightarrow{r}_{i'}, E_{j'}) = \frac{\sum_{k'=1}^{K'} AF(i', j', k')}{H \Delta E_{j'}}, \tag{8.31}$$

The benefit of the analytical estimator is that each scattered or born particle (source) results in a tally. Therefore, in a highly scattering medium, one may achieve a very precise result after a few thousand histories without using variance reduction techniques. This result can be very inaccurate given that most of the sampling is performed far from the point of interest. Further, the technique may lead to very small weights (contributions) if the particle site is far from the point of interest or if the probability of scattering into the solid angle of interest is very small. The opposite situation exists near the point of interest, where $\frac{1}{s^2}$ becomes very large and, consequently, may bias the results. To avoid this difficulty, we consider an exclusion volume around the point of interest and perform analog Monte Carlo within that volume. The size of the exclusion volume must be decided by experimentation. The MCNP code offers a similar tallying option referred to as DXTRAN [96]; often the exclusion volume is represented by a spherical region around the point of interest; in MCNP this volume is referred to as *DXTRAN spheres*.

8.4   TIME-DEPENDENT TALLYING

Incorporation of time-dependency in a Monte Carlo simulation is rather simple and mainly requires additional bookkeeping.

Suppose that a particle is born at time t_0 with speed v_0. If the particle travels a distance s_0 to its first interaction, then the interaction time is estimated by

$$t_1 = t_0 + \frac{s_0}{v_0} \tag{8.32}$$

After n interactions, if the particle crosses an interface, as shown in Figure 8.4, the time at which the $(n + 1)^{th}$ interaction occurs is

$$t_{n+1} = t_n + \frac{s}{v} \tag{8.33}$$

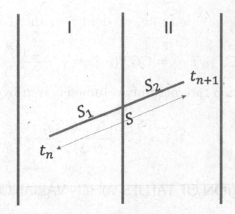

Figure 8.4 Particle crossing an interface

Therefore, the time that particle crosses the interface is given by

$$t = t_n + (t_{n+1} - t_n)\frac{s_1}{s} \qquad (8.34)$$

Note that as long as the particle speed is $\leq 0.2c$, where c is the speed of light, we can use the classical mechanics formulations.

All the tallying techniques, i.e., estimators discussed in Section 8.3 can be extended easily to account for time-dependency. For each estimator, a new counter should be defined to tally particles over time intervals. Here, we only derive/discuss new formulations for the collision estimator, the other estimators may be treated similarly.

For the collision estimator, we define a time-dependent counter CFT as

$$CFT(i, j, k, n) = CFT(i, j, k, n) + \frac{w}{\Sigma_t(E)}, \qquad (8.35)$$

which accumulates the weight w of all particles with energy E within ΔE_j, moving in direction $\hat{\Omega}$ within $\Delta\Omega_k$, that suffer a collision within ΔV_i in a time interval of $\Delta t_n (= t_{n+1} - t_n)$. Then, the time-dependent angular flux formulation is expressed by

$$\psi(\vec{r}_i, E_j, \hat{\Omega}_k, t_n) = \frac{CFT(i, j, k, n)}{H\Delta V_i \Delta E_j \Delta\Omega_k} \qquad (8.36)$$

and, the time-dependent scalar flux is given by

$$\phi(\vec{r}_i, E_j, t_n) = \frac{\sum_{k=1}^{K} CFT(i, j, k, n)}{H\Delta V_i \Delta E_j} \qquad (8.37)$$

To estimate the collision rate, we define another counter as

$$CCT(i, j, n) = CCT(i, j, n) + w \frac{\Sigma_c(E)}{\Sigma_t(E)} \tag{8.38}$$

then, the reaction rate per unit volume is given by

$$R(\overrightarrow{r}_i, t_n) = \frac{\sum_{j=1}^{J} CCT(i, j, n)}{H \Delta V_i} \tag{8.39}$$

8.5 FORMULATION OF TALLIES WHEN VARIANCE REDUCTION USED

In analog Monte Carlo, each particle is born with weight one, and, if tallied, it is counted with weight one. However, this is not true when variance reduction in used in a Monte Carlo simulation in which a particle weight is adjusted after every biased event. This means a particle may contribute several tallies (x) to a sample average with different weights (w), and, therefore, particle tally contribution x_h at each history h is expressed by

$$x_h = \sum_{\ell=1}^{n_h} w_{h,\ell} x_{h,\ell} \tag{8.40}$$

where n_h refers to the number of events that contribute to the sample average in the hth history, $x_{h,\ell}$ refers to the tally for ℓ^{th} event in h^{th} history, and $w_{h,\ell}$ refers to particle weight at ℓ^{th} event in h^{th} history. Using the above formulation for contribution of each history, the sample average after H histories is expressed by

$$\overline{x} = \frac{1}{H} \sum_{h=1}^{n_h} x_h = \frac{1}{H} \sum_{h=1}^{H} \sum_{\ell=1}^{n_h} w_{h,\ell} x_{h,\ell} \tag{8.41}$$

Now, it is constructive to use the above formulation for sample averages of different quantities, such as flux, current, and reaction rates introduced in Sections 8.3 and 8.4. For example, to determine the angular flux via collision estimator using Equation 8.10, the tally is expressed by

$$x_{h,\ell} = \frac{1}{\Delta V_i \Delta E_j \Delta \Omega_k} \left(\frac{1}{\Sigma_t(E_{h,\ell})} \right) \tag{8.42}$$

The above equation yield the tally per unit volume, unit energy, unit steradian due to collisions within ΔV_i, at energy $E_{h,\ell}$ within ΔE_j, along direction $\hat{\Omega}$ within $\Delta\Omega_k$. Then, the average angular flux is given by

$$\overline{\psi_{i,j,k}} = \frac{1}{H\Delta V_i \Delta E_j \Delta\Omega_k} \sum_{h=1}^{H} \sum_{\ell=1}^{n_h} \frac{w_{h,\ell}}{\Sigma_t(E_{h,\ell})} \quad (8.43)$$

Using equations of tallies for different quantities using the different estimators given in Sections 8.3 and 8.4, we can obtain new equations that include the effect of different contributing events during the particle history similar to Equation 8.43.

8.6 ESTIMATION OF RELATIVE UNCERTAINTY OF TALLIES

In previous sections, we presented different formulations for tallying particle flux, current, and reaction rates in finite ranges of space, energy, and/or direction. Here, we will elaborate on the estimation of uncertainties for these quantities.

The distinction between x_h and the $x_{h,\ell}$'s discussed in Section 8.5 is very important when calculating the uncertainty of tally x. This is because, while the x_h's are independent events, the $x_{h,\ell}$'s within an individual history are not. Note that Equation (4.123) requires that the x_i's are independent events; therefore, to use Equation (4.123) properly, the x_i's must be the x_h's, i.e., complete histories, not the individual event $x_{h,\ell}$'s.

Using Equation (4.123), we may write formulations for variances of different quantities. For example, from the angular flux expressed by Equation 8.43, the sample relative uncertainty for angular flux after H histories is given by

$$R_{\overline{\psi}} = \sqrt{\frac{1}{H-1}\left(\frac{\overline{\psi^2}}{\overline{\psi}^2} - 1\right)} \quad (8.44)$$

where

$$\overline{\psi^2}_{i,j,k} = \frac{1}{H}\sum_{h=1}^{H}\left(\sum_{\ell=1}^{n_h} \frac{w_{h,\ell}}{\Delta V_i \Delta E_j \Delta\Omega_k \Sigma_t(E_{h,\ell})}\right)^2 \quad (8.45)$$

Similar to the above equation, one can derive formulations for the relative uncertainty for other quantities. Note that to estimate the uncertainty associated with a detector count, one can consider using the formulations derived in Section 4.6.

8.7 UNCERTAINTY IN A RANDOM VARIABLE DEPENDENT ON OTHER RANDOM VARIABLES

In the previous section, the relative uncertainty formulations were derived for a sampled random variable; however, there are situations that a random variable (x) depends on other random variables, i.e., $u_1, u_2, u_3, ..$, that is,

$$x \equiv x(u_1, u_2, u_3, \cdots \cdots) \tag{8.46}$$

In such a situation, the variance of x is obtained by employing the *propagation* of uncertainty formulation

$$\sigma_x^2 = \sum_{i=1}^{N} \sum_{j=1}^{N} \frac{\partial x}{\partial u_i} \frac{\partial x}{\partial u_j} \sigma_{u_i u_j}^2 \tag{8.47}$$

where, $\sigma_{u_i u_i}^2 = \sigma_{u_i}^2$.

Example 8.1
If a random variable (z) is obtained by a linear combination of two other random variables given by

$$z = x + y \tag{8.48}$$

Using Equation 8.47, then the variance for the random variable z is derived as follows

$$\sigma_z^2 = \frac{\partial z}{\partial u_1} \left(\frac{\partial z}{\partial u_1} \sigma_{u_1, u_1}^2 + \frac{\partial z}{\partial u_2} \sigma_{u_1, u_2}^2 \right) + \frac{\partial z}{\partial u_2} \left(\frac{\partial z}{\partial u_1} \sigma_{u_1, u_2}^2 + \frac{\partial z}{\partial u_2} \sigma_{u_2, u_2}^2 \right)$$

or

$$\sigma_z^2 = \left(\frac{\partial z}{\partial u_1} \right)^2 \sigma_{u_1, u_1}^2 + 2 \frac{\partial z}{\partial u_1} \frac{\partial z}{\partial u_2} \sigma_{u_1, u_2}^2 + \left(\frac{\partial z}{\partial u_2} \right)^2 \sigma_{u_2, u_2}^2 \tag{8.49}$$

Considering $u_1 = x$ and $u_2 = y$, then the formulation of variance of z given by

$$\sigma_z^2 = \sigma_x^2 + 2\sigma_{x,y}^2 + \sigma_y^2 \tag{8.50}$$

Hence, the sample variance for random variable z is given by

$$S_z^2 = S_x^2 + 2S_{x,y}^2 + S_y^2 \tag{8.51}$$

where

$$S_x^2 = \frac{1}{N-1} \sum_{i=1}^{N} (x_i - \overline{x})^2$$

$$S_y^2 = \frac{1}{N-1} \sum_{i=1}^{N} (y_i - \overline{y})^2 \qquad (8.52)$$

$$S_{x,y}^2 = \frac{1}{N-1} \sum_{i=1}^{N} (x_i - \overline{x})(y_i - \overline{y})$$

Example 8.2

The random variable z is a function of random variables, x and y, as follows

$$z = \frac{x}{y} \qquad (8.53)$$

then, the variance of $\overline{z}$ is given by

$$S_{\overline{z}}^2 = \left(\frac{\overline{x}}{\overline{y}}\right)^2 \left(\frac{S_x^2}{\overline{x}^2} - 2\frac{S_{x,y}^2}{\overline{x}\,\overline{y}} + \frac{S_y^2}{\overline{y}^2}\right) \qquad (8.54)$$

8.8 REMARKS

In this chapter, we discussed different techniques for tallying particles in a Monte Carlo particle transport simulation. For each technique, we have developed counters and the corresponding formulations for estimating physical quantities such as particle flux, current, and reaction rates. We have demonstrated that considering time-dependency in a Monte Carlo simulation is only a matter of bookkeeping and increased need for computer resources (memory and time). It is important to note that selecting an appropriate tallying technique may have a significant impact on achieving a precise and accurate result in a reasonable amount of time. Finally, formulations for tally uncertainties and uncertainty propagation are derived and discussed.

PROBLEMS

1. Starting with the program from Problem 1, Chapter 7, implement flux tallies using the collision estimator and path length estimator. The variances of the tallies can be calculated using

Equation (4.122). Divide the shield into the following number of regions for tallying.

a. 10 regions

b. 50 regions

Use the following parameters: $\Sigma_s/\Sigma_t = 0.8$, $\theta = 0$, $\Sigma_t d = 8$. Stop the simulation when the highest relative tally error is less than 10%, or 10,000,000 (10^7) particles have been simulated, whichever comes first. Plot the fluxes, relative uncertainties, and FOMs (Equation 5.10). Explain and comment on the results.

2. Repeat Problem 1, but use geometric splitting with five importance regions (of importance 1, 2, 4, 8, and 16). Note that there are five regions for importance, but you should use the same tally regions as Problem 1. Compare these results to those without splitting.

3. Modify the program you wrote for Problem 3, Chapter 7 to tally the average scalar flux per unit mean free path. Test your program based on a 1-D slab of size 10 cm, with a purely absorbing material of $\Sigma_t = \Sigma_a = 0.2 \ cm^{-1}$, and a planar mono-directional $(\mu = 1)$ source placed on its right boundary. Consider both collision and path-length estimator techniques, and compare your results to analytic results.

4. Modify Problem 3 to be able to use the surface-crossing technique for determination of scalar flux. Compare your results with those obtained based on collision and path-length estimators.

5. Examine the effect of cell size for estimating the scalar flux based on the collision, path-length, and surface-crossing techniques with a maximum relative error of 5%. Vary the cell size from 10% of *mfp* to 1 *mfp*.

6. If a random variable is obtained as a weighted average of two estimated random variables (which represent the same physical quantity), that is:

$$z = \alpha x + (1 - \alpha)y$$

Determine the optimal α by minimizing the variance of z.

CHAPTER 9

Geometry and particle tracking

CONTENTS

9.1   INTRODUCTION

Geometric modeling is one of the most important features of a Monte Carlo algorithm, because it can significantly impact areas of application, input preparation, degree of accuracy, and computation time. Throughout the years, different groups have introduced different flavors of geometric algorithms depending on their needs and limitations. Wang [114] provides a good review of these algorithms. More commonly used approaches include:

- Combinatorial geometry, which forms a model by combining (using Boolean algebra) simple primitive objects, e.g., cuboids, cylinders, ellipsoids. This approach limits the user to modeling of simple or idealized objects.

- The use of voxelized or triangular meshing for generation of a model. This approach, which is commonly used in medical applications, has been implemented into codes, such as GEANT [1]

and PENELOPE [89]. This technique may have low resolution
and difficulty in modeling deformed bodies because of the mesh
size.

- Employment of a standard computer-aided design (CAD) package, which generally may result in significant overhead. Two
  approaches are considered: (i) Converter approach in which an
  auxiliary software translates the CAD representation to the
  geometry input of a Monte Carlo code of interest; and (ii) CAD-
  base Monte Carlo in which CAD software is coupled with a standard Monte Carlo (MC) code such that geometric-related tasks
  are performed within CAD, while all the other tasks are performed with the standard MC code [114]. Although the second
  approach provides significant flexibility in modeling highly complex problems, it is plagued by high computational cost. For
  example, the method used by Wang has shown an increase in
  computational cost by a factor of three compared to standard
  codes.

- Use of more flexible combinatorial geometry that defines surfaces,
  instead of objects, which, in turn, are combined via Boolean algebra to make cells and eventually the whole geometric model. This
  approach is used in the MCNP (Monte Carlo Neutron and Photon) code system [96].

Irrespective of the approach, the geometric model can significantly
impact accuracy and computation time of a Monte Carlo simulation,
and users should avoid unnecessary detail or the use of convoluted logic
in making a model. In the rest of this chapter, we will elaborate on the
last approach as implemented in MCNP.

9.2 COMBINATORIAL GEOMETRY APPROACH

A combinatorial geometry approach describes surfaces and Boolean
combinations of surfaces define cells. First, the surfaces are defined.
Then, the surfaces are used to construct cells, and cells are used with
Boolean logic to define other complicated cells. The Boolean operators are ADD, OR, and NOT. In the combinatorial geometry commu-
nity, the three operators are referred as $intersection(\cap)$, $union(\cup)$, and
$complement(\#)$, respectively. The intersection and union operators are

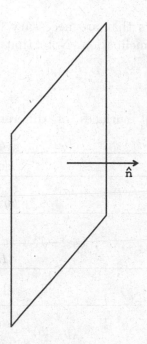

Figure 9.1 Schematic of a planar surface and its normal unit vector

used with surfaces to make cells and the complement operator is used to make the complement of a cell.

9.2.1 Definition of a surface

Here, the analytical formulations of different orders are needed in order to represent cells of different shapes. For example, the formulation for a planar surface is expressed by

$$f(x, y, z) = ax + by + cz + d = 0 \tag{9.1}$$

Each surface has two senses: positive and negative. Commonly, the positive sense is chosen along the normal unit vector to the surface as shown in Figure 9.1.

Equation 9.1 provides information on the points on the surface, while the points on the positive side of the surface are obtained when $f(x, y, z) > 0$, and points to the negative side of the surface are obtained when $f(x, y, z) < 0$.

Table 9.1 provides analytical formulations for a set of surfaces (first to fourth order) as implemented in the MCNP code. Each surface has a set of free parameters that are necessary for variable positioning, size, and/or degree of inclination. Note that the planar surfaces are infinite.

Table 9.1 Equations of surfaces of different orders (1-4) with parameters[*]

Type	Equation
Plane	$ax + by + cz + d = 0$
Sphere	$(x-a)^2 + (y-b)^2 + (z-c)^2 - R^2 = 0$
Cylinder	
Parallel to $x - axis$	$(y-b)^2 + (z-c)^2 - R^2 = 0$
Parallel to $y - axis$	$(x-a)^2 + (z-c)^2 - R^2 = 0$
Parallel to $z - axis$	$(x-a)^2 + (y-b)^2 - R^2 = 0$
Cone	
Parallel to $x - axis$	$\sqrt{(y-b)^2 + (z-c)^2} - t(x-a) = 0$
Parallel to $y - axis$	$\sqrt{(x-a)^2 + (z-c)^2} - t(y-b) = 0$
Parallel to $z - axis$	$\sqrt{(x-a)^2 + (y-b)^2} - t(z-c) = 0$
General Ellipsoid hyperboloid, or paraboloid	$b(x-a)^2 + d(y-c)^2 + e(z-f)^2 + 2g(x-h) + 2i(y-j) + 2k(z-\ell) + m = 0$
Torus (elliptic & circular)	
Parallel to $x - axis$	$\frac{(x-a)^2}{B^2} + \frac{(\sqrt{(y-b)^2+(z-c)^2}-A)^2}{C^2} - 1 = 0$
Parallel to $y - axis$	$\frac{(y-b)^2}{B^2} + \frac{(\sqrt{(x-a)^2+(z-c)^2}-A)^2}{C^2} - 1 = 0$
Parallel to $z - axis$	$\frac{(z-c)^2}{B^2} + \frac{(\sqrt{(x-a)^2+(y-b)^2}-A)^2}{C^2} - 1 = 0$

[*] MCNP manual [96]

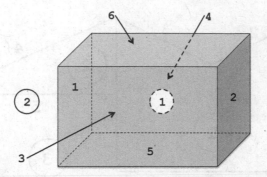

Figure 9.2 Schematic of a cell (parallelpiped) made of six planar surfaces

9.2.2 Definition of cells

Cells or material regions are formed by applying Boolean operations to surfaces and other cells. More specifically, a cell is formed by *intersection* and *union* operations on the positive/negative sides of its bounding surfaces, and by *complement* operation on other cells. To illustrate this, let's make a parallelepiped by using six planar surfaces (1 to 6) as shown in Figure 9.2.

Utilizing Boolean operations, we may form the parallelepiped (cell 1) by *intersection* of specific side of each surface as follows

$$Cell\ 1 : +1 \cap -2 \cap +3 \cap -4 \cap +5 \cap -6 \tag{9.2}$$

Of course, more complicated physical models can be made by combining higher-order surfaces.

9.2.3 Examples for irregular cells

Here, we consider a 2-D diagram shown in Figure 9.3, and write the necessary equations describing cells 1 to 3. As shown in the figure, we have assigned ID's for all the surfaces from 1 to 7, and these surfaces to make the cells as follows:

- Cell 1 is somewhat complicated as it include two inclined surfaces (1 & 2) which make a concave corner into the cell. The equation for cell is given by

$$+1 \cap (-2 \cup -3) \cap -7 \cap +8 \tag{9.3}$$

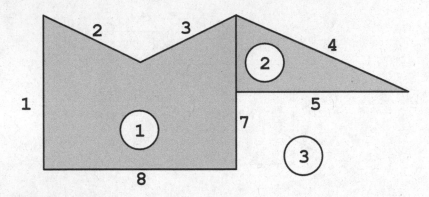

Figure 9.3 Schematic of a example with irregular shapes

- Cell 2 is quite simple, and can be described by

$$+7 \cap -4 \cap +5 \qquad (9.4)$$

- Cell 3 is more complicated because it includes two corners, one made by surfaces 2 and 3, and the other by surfaces 5 and 7. This cell is expressed by

$$-1 \cup (+2 \cap +3) \cup +4 \cup (+5 \cap +7) \cup -8 \qquad (9.5)$$

Note that the parentheses in the above equations indicate order of operation. This means that the operations between 2 and 3, as well as between −5 and 7 are performed before adding these to the other regions. It is important to note that commonly cells can be described in different ways; for example, cell 3 can be made by a combination of complements of cells 1 and 2 as follows

$$\#1 \cap \#2 \qquad (9.6)$$

It is generally better not to use convoluted logic, e.g., different nested operations, as particle tracking through the problem geometry represent a major fraction of computational cost of a Monte Carlo simulation.

9.3 DESCRIPTION OF BOUNDARY CONDITIONS

Any physical problem has a finite size, which is identified by the boundary surfaces. Boundary conditions (*BCs*) provide information on the

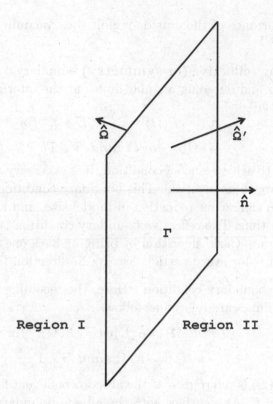

Region I   **Region II**

Figure 9.4 Schematic of an interfacial boundary between two regions

behavior of a particle angular flux at the boundary of a calculation model. To describe different boundary conditions, we will use Figure 9.4, which depicts an interfacial boundary surface (Γ) between regions I and II, and unit vectors $\hat{\Omega}$ and $\hat{\Omega}'$ referring to particle directions entering regions I and II, respectively, and $\hat{n}$ is unit vector normal to surface Γ. We will discuss five boundary conditions including *vacuum, specular reflective, albedo, white, and periodic.*

- **Vacuum** boundary condition - If we consider region II in Figure 9.4 is void (or vacuum), then no particles will be reflected back into region I. Hence, the incoming angular flux to region is expressed by

$$\psi(\overrightarrow{r}, E, \hat{\Omega}) = 0, \; for \; \hat{n} \cdot \hat{\Omega} < 0 \; and \; \overrightarrow{r} \epsilon \; \Gamma \qquad (9.7)$$

  where $\overrightarrow{r}$ is particle position on the boundary, and E and $\hat{\Omega}$ are energy and direction of the particle. The vacuum boundary condition can be incorporated into a Monte Carlo algorithm by setting the importance of the outside region (i.e., vacuum region) equal to zero.

- **Specular reflective (or symmetry)** boundary condition - The incoming and outgoing angular fluxes at the interface are equal, i.e.

$$\psi(\overrightarrow{r}, E, \hat{\Omega}) = \psi(\overrightarrow{r}, E, \hat{\Omega}), \ \ for$$
$$\hat{n} \cdot \hat{\Omega} = -\hat{n} \cdot \hat{\Omega}^{'} \ and \ \overrightarrow{r} \epsilon \ \Gamma$$

(9.8)

In order to achieve such a condition, it is necessary that regions I and II are to be identical. This boundary condition if applicable results in significant reduction in model size, and therefore computation time. The reflective boundary condition is incorporated into a Monte Carlo algorithm by bringing back one particle along direction $\hat{\Omega}$ for every particle leaving in direction $\hat{\Omega}^{'}$.

- **Albedo** boundary condition - Here, the incoming and outgoing angular fluxes are related as follows

$$\psi(\overrightarrow{r}, E, \hat{\Omega}) = \alpha(E)\psi(\overrightarrow{r}, E, \hat{\Omega}), \ \ for$$
$$\hat{n} \cdot \hat{\Omega} = -\hat{n} \cdot \hat{\Omega}^{'} \ and \ \overrightarrow{r} \epsilon \ \Gamma$$

(9.9)

where $\alpha(E)$ is referred to as the albedo coefficient for the particle of energy E. At a surface with the albedo boundary condition, a fraction $(\alpha(E))$ of particles leaving the surface, e.g., along $\hat{\Omega}^{'}$ into region II, will be reflected back into the region I. This boundary condition is used to avoid modeling a region, while still keeping its impact, i.e., the reflection of some fraction of particles.

The albedo boundary condition is incorporated into a Monte Carlo algorithm by reflecting back along direction $\hat{\Omega}$ a fraction of particles entering region II along direction $\hat{\Omega}^{'}$.

- **White** boundary condition - particles leaving region I through a white boundary are reflected isotropically back to the region. The formulation for the white boundary condition is given by

$$\psi(\overrightarrow{r}, E, \hat{\Omega}) = \frac{\int_{2\pi+} d\Omega' \hat{n} \cdot \hat{\Omega}^{'} \psi(\overrightarrow{r}, E, \hat{\Omega}^{'})}{\int_{2\pi+} d\Omega' \hat{n} \cdot \hat{\Omega}^{'}}$$
$$\hat{n} \cdot \hat{\Omega} < 0 \ and \ \overrightarrow{r} \epsilon \ \Gamma$$

(9.10)

To implement the white boundary condition, any particle that intersects a white boundary should be reflected back with a cosine distribution ($p(\mu) = \mu$).

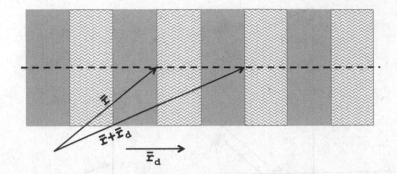

Figure 9.5 Schematic of an example for periodic boundary condition

- **Periodic** boundary condition - In problems with physical periodicity, such as fuel assemblies or fuel cells in a reactor, in special cases of an infinite system, one may be able to establish that angular flux distribution on one boundary($\vec{r}$)is equal to the angular distribution on another boundary ($\vec{r} + \vec{r_d}$) in a periodic manner, as shown in Figure 9.5. Mathematically, the periodic boundary condition is expressed by

$$\psi(\vec{r} + \vec{r_d}, E, \hat{\Omega}) = \psi(\vec{r}, E, \hat{\Omega}) \qquad (9.11)$$

To incorporate this boundary condition, a particle entering boundary at $\vec{r}$ will reenter at its periodic surface at $\vec{r} + \vec{r_d}$.

9.4   PARTICLE TRACKING

In addition to setting up the geometric model and its boundary conditions, it is necessary to determine the location of each particle within the physical model. This is accomplished by examining the position of each particle relative to interfacial and outside boundaries. For example, as depicted in Figure 9.6, to determine the position (x_0, y_0, z_0) of a particle relative to a surface defined by

$$f(x, y, z) = 0 \qquad (9.12)$$

we need to substitute particle position into the surface equation, and compare with zero as follows

- If $f(x_0, y_0, z_0) < 0$, particle is inside the medium

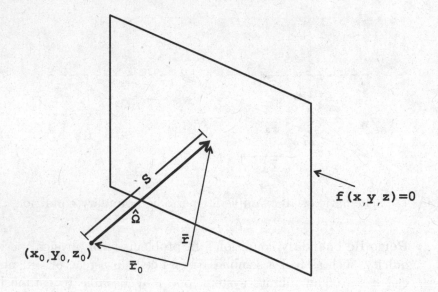

Figure 9.6 Schematic of particle tracking to a boundary

- If $f(x_0, y_0, z_0) = 0$, particle is on the surface

- If $f(x_0, y_0, z_0) > 0$, particle is outside the medium

Further, it is necessary to determine the intersection point of the particle path with the surface, as shown in Figure 9.6. This means that we have to determine (x, y, z) corresponding vector position $\vec{r}$ expressed by

$$\vec{r} = \vec{r_0} + s\hat{\Omega} \tag{9.13}$$

The components (x, y, and z) of vector position $\vec{r}$ are obtained by finding its projections along x, y, and z axes as follows

$$
\begin{aligned}
x &= \hat{i} \cdot \vec{r} = \hat{i} \cdot \vec{r_0} + \hat{i} \cdot \hat{\Omega} = x_0 + su \\
y &= \hat{j} \cdot \vec{r} = \hat{j} \cdot \vec{r_0} + \hat{j} \cdot \hat{\Omega} = y_0 + sv \\
z &= \hat{k} \cdot \vec{r} = \hat{k} \cdot \vec{r_0} + \hat{k} \cdot \hat{\Omega} = z_0 + sw
\end{aligned}
\tag{9.14}
$$

where u, v, and w are direction cosines of angles between $\hat{\Omega}$ and x, y, and z axes, respectively. To determine s (path length between particle position and surface) x, y, and z from Equation 9.14 have to be substituted in the surface equation (9.12), i.e.

$$f(x_0 + su, y_0 + su, z_0 + sw) = 0 \tag{9.15}$$

Note that in case of higher order surfaces, the path-length should be set to the smallest positive root of above equation.

9.5 REMARKS

This chapter discussed different techniques for the generation of a model for a Monte Carlo algorithm, and elaborated on the combinatorial geometry approach. It is noted that the technique used for defining the geometry can impact the flexibility and cost of a Monte Carlo simulation. A discussion on how the Boolean operators are employed in a combinatorial geometry technique to create arbitrary objects was presented. Through a few simple examples, the use and difficulties of the technique are demonstrated. Finally, different boundary conditions in particle transport and particle tracking are discussed. As a final note, since tracking is one of the most time-consuming parts of a Monte Carlo simulation, efficient algorithms should be devised and complicated combinations of surfaces and cells should be avoided.

PROBLEMS

1. Write Boolean instructions for making cells 1 and 2 (identified by circles) in Figure 9.7.

2. Write the Boolean instructions for making the cells 1-3 in Figure 9.8.

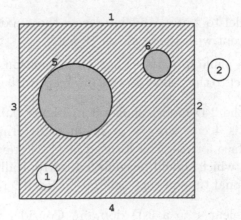

Figure 9.7 Problem 1

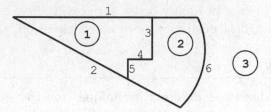

Figure 9.8 Problem 2

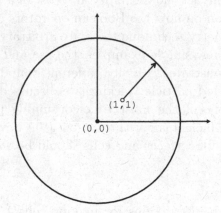

Figure 9.9 Problem 4

3. Determine the parameters a, b, c, and d for a planar surface as given in Table 9.1 for the following conditions:

   a. Parallel to z-axis, and intersects the x-axis at 60°, and its intercept with the y-axis is 10 cm.

   b. If the plane in part (a) has an inclination angle of 30°with respect to the z-axis, and its intercept with z-axis is at 10 cm.

4. Consider the 2-D model depicted in Figure 9.9. If the radius of the circle is 4 cm, write a program to determine the distance to the surface for particles located at $(1,1)$ traveling in different directions, which should be sampled isotropically. Determine the computational time of this calculation for 10^5 particles.

5. Move Problem 4 to a 3-D domain. Consider a spherical shell with a radius of 4 cm and particles positioned at $(1,1,1)$. Again, write a program to sample particle direction isotropically and

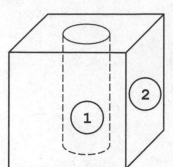

Figure 9.10 Problem 8

evaluate the particle distance to the spherical shell. Determine the computation cost of this calculation for 10^5 particles.

6. Replace the spherical shell from Problem 5 with ellipsoidal and toroidal shells, and perform a similar timing analysis for these surfaces. For the ellipsoid, consider x, y, and z radii of 3, 4 and 5 cm, respectively. For the torus, consider a major radius of 2 cm and minor radius of 1 cm. Both of them are centered about the origin.

7. Derive probability density functions for uniform sampling of source particles from a cylinder, sphere, and cube.

8. Consider a cylinder inside a cube placed in a vacuum as depicted in Figure 9.10. An isotropic line source is placed at the center of the cylinder. Considering that both the cylinder and the cube are void:

    a. Develop the necessary formulation for sampling the pathlength a source particle travels in the cylinder and the cube.

    b. Write a computer code to determine the total path-length of the source particles in both the cylinder and the cube.

    c. Determine the volumes of regions 1 (cylinder) and 2 (outside the cylinder).

9. Modify the program developed in Problem 8 to calculate the average distance that a source particle must travel before re-entering the cylinder for:

    a. A specular reflective boundary condition.

    b. A white boundary condition.

CHAPTER 10

Eigenvalue (criticality) Monte Carlo method for particle transport

CONTENTS

10.1   INTRODUCTION

Thus far, only the fixed-source Monte Carlo particle transport (with application to areas of radiation shielding, dosimetry, and nuclear security) has been discussed. In this chapter, we will introduce the Monte Carlo method as applied to eigenvalue problems that are encountered in reactor physics as well as nuclear safeguards and nonproliferation. The major difference between the two problem types is that, in the eigenvalue problems, source distribution is unknown and, therefore, there is a need for an extra layer of computation to obtain the source distribution.

The most common approach for solving an eigenvalue neutron transport problem is the power-iteration technique in which the fission neutron source (or power) is iterated on until it converges within a prescribed tolerance [65, 4]. In the Monte Carlo method, this technique is implemented by introducing a set of user-defined parameters that vary for different problems, and generally require experimentation and observation for selecting a "correct" combination of parameters that yields a converged source distribution and precise problem solutions. This situation becomes more complicated for problems that have a high dominance ratio or are loosely coupled (these terms will be described in further detail later). For such problems, the power-iteration technique may lead to biased and/or erroneous results [38, 28, 17, 101, 60]. Significant efforts have been devoted to the development of diagnostic techniques for detection of source convergence [99, 119, 117, 5, 6], and for development of alternative techniques for eigenvalue Monte Carlo calculations [7, 27, 92, 118, 116].

In this chapter, first we will introduce the theory of the power-iteration technique for solving eigenvalue problems. Second, we will develop the power-iteration as implemented for eigenvalue Monte Carlo particle transport simulation, and elaborate on issues/shortcomings of this method. Third, we will discuss the concept and methods for detection of source convergence. Finally, we will demonstrate the shortcomings of the standard eigenvalue Monte Carlo simulation, and argue the need for alternative techniques.

10.2 THEORY OF POWER ITERATION FOR EIGENVALUE PROBLEMS

The eigenvalue linear Boltzmann (or transport) equation within a phase space $(d^3 r dE d\Omega)$ is expressed by

$$H\psi = \frac{1}{k}F\psi, \qquad in \ V$$
$$\psi = \tilde{\psi}, \qquad for \ \hat{n} \cdot \hat{\Omega}, \ \overrightarrow{r} \in \Gamma \tag{10.1}$$

where $\psi \equiv \psi(\overrightarrow{r}, E, \hat{\Omega})$ is the eigenfunction (angular flux) at position $\overrightarrow{r}$ within $d^3 r$, at direction $\hat{\Omega}$ within $d\Omega$, and with energy E within dE, k is the eigenvalue (multiplication factor), V refers to volume, Γ refers to surface area, $\tilde{\psi}$ refers to a given boundary value, and operators H and F are expressed by

$$H = \hat{\Omega} \cdot \nabla + \Sigma_t(\overrightarrow{r}, E) - \int_0^\infty dE' \int_{4\pi} d\Omega' \Sigma_s(\overrightarrow{r}, E' \to E, \mu_0)$$

and

$$F = \frac{\chi(E)}{4\pi} \int_0^\infty dE' \int_{4\pi} d\Omega' \nu \Sigma_f(\overrightarrow{r}, E') \tag{10.2}$$

Where, Σ_t is the total cross-section, $\Sigma_s(\overrightarrow{r}, E' \to E, \mu_0)$ is the differential scattering cross-section, Σ_f is the fission cross-section, χ is the fission neutron spectrum, and ν is the average number of fission neutrons generated per fission.

To solve for the eigenfunction (ψ), we rewrite Equation 10.1 as follows

$$\psi = \frac{1}{k}(H^{-1}F)\psi = \frac{1}{k}M\psi \tag{10.3}$$

where, $M = H^{-1}F$.

In principle, the operator M has several eigenvalues (k_i) and corresponding eigenfunctions (ψ_i). We are interested in solving for the fundamental values, i.e., k_0 and ψ_0, as k_0 is the largest eigenvalue, i.e.,

$$k_0 > |k_1| > |k_2| > \cdots \tag{10.4}$$

And therefore a general solution can be expressed by in terms of eigenfunctions as follows

$$u = \sum_0^\infty a_i u_i \tag{10.5}$$

This approach, however, is not practical for a general problem. Since the source in Equation 10.3 is unknown, it is necessary to use an iterative approach in which the right-hand side of Equation 10.1, i.e., source is obtained from a previous iteration. This means after n iteration, Equation 10.3 reduces to

$$\psi^{(n)} = \frac{1}{k^{(n-1)}} M \psi^{(n-1)}, \quad n = 1, 2, \cdots \tag{10.6}$$

The above equation is used to obtain the nth eigenfunction, and the corresponding eigenvalue is obtained by

$$k^{(n)} = \frac{<M\psi^{(n)}>}{<M\psi^{(n-1)}>} = \frac{<M\psi^{(n)}>}{k^{(n)} <\psi^{(n)}>} \tag{10.7}$$

where Dirac signs ($<>$) refer to integration over all independent variables. This iteration process is referred to as the power (or source) iteration. Generally, the process is terminated if the eigenfunction distribution and eigenvalue are converged within given tolerances.

To examine the convergence behavior of the fundamental eigenfunction, we consider that the *zeroth-iteration* eigenfunction is given in terms of the normalized eigenfunctions (u_i's) as follows:

$$\psi^0 = \sum_i a_i u_i. \tag{10.8}$$

Then, we obtain i^{th} iteration eigenfunction, $\psi^{(i)}$ for $i = 0, n$ using the following equations:

$$\psi^{(1)} = \frac{1}{k^{(0)}} M \psi^{(0)}$$

$$\psi^{(2)} = \frac{1}{k^{(1)}} M \psi^{(1)} = \frac{1}{k^{(0)} k^{(1)}} M^2 \psi^{(0)}$$

$$\cdots$$

$$\cdots \tag{10.9}$$

$$\cdots$$

$$\psi^{(n)} = \frac{1}{\alpha} M^n \psi^{(0)}$$

where $\alpha = \prod_{i=0}^{n-1} k^{(i)}$.

Now, if we substitute Equation 10.8 for $\psi^{(0)}$ in Equation 10.9, we obtain

$$\psi^{(n)} = \frac{1}{\alpha} M^n \sum_i a_i u_i = \sum_i \frac{a_i}{\alpha} M^n u_i \tag{10.10}$$

Considering Equation 10.6 for each eigenfunction, Equation 10.10 reduces to

$$\psi^{(n)} = \sum_i \frac{a_i}{\alpha} k_i^n u_i \tag{10.11}$$

Now, if we divide Equation 10.11 by k_0^n, we obtain the following equation

$$\frac{\psi^{(n)}}{k_0^n} = \frac{a_0}{\alpha} u_0 + \frac{a_1}{\alpha} \left(\frac{k_1}{k_0} \right)^n u_1 + \frac{a_2}{\alpha} \left(\frac{k_2}{k_0} \right)^n u_2 + \cdots \tag{10.12}$$

To examine the convergence behavior of Equation 10.12, we divide the inequality Equation 10.4 by k_0 to obtain

$$1 > \left| \frac{k_1}{k_0} \right| > \left| \frac{k_2}{k_0} \right| > \cdots \tag{10.13}$$

This means that $\left| \frac{k_1}{k_0} \right|$ ratio, referred to as the *dominance ratio*, is the largest term in the series. Hence, the convergence of Equation 10.12 to the fundamental eigenfunction is dependent on the value of the *dominance ratio*. If the ratio is close to 1, a situation referred to as *High Dominance Ratio* (HDR), the convergence to the fundamental eigenfuction becomes very slow. Accordingly, a Monte Carlo eigenvalue calculation will have difficulties in achieving a converged solution for problems which are plagued by HDR.

10.3 MONTE CARLO EIGENVALUE CALCULATION

As indicated, the main difference between eigenvalue and fixed-source problems is the fact that the source is unknown. Hence, an initial particle source distribution and corresponding eigenvalue have to be guessed. Further, since the particle source is generated through the fission process, it is necessary to discuss and derive the necessary formulations for sampling the fission neutrons, i.e., their count, energies, and directions. Section 10.3.1 derives the necessary formulations for sampling fission neutrons, and Section 10.3.2 discusses a procedure for performing Monte Carlo eigenvalue simulations based on the power-iteration technique. Section 10.3.3 discusses different estimators for sampling fission neutrons. Section 10.3.4 provides a methodology for combining different estimators.

10.3.1   Random variables for sampling fission neutrons

The byproducts of a fission process commonly are two fission products and several charged and neutral particles. In an eigenvalue problem, the fission neutrons are the necessary source for sustaining the *fission chain*; henceforth, this section is devoted to the sampling of the *fission neutrons*. Similar formulations can be derived for other byproducts of a fission process.

There are three random variables associated with the fission neutrons: (1) number, (2) energy, and (3) direction. The following subsections are devoted to the derivation of the FFMCs corresponding to these random variables.

10.3.1.1   Number of fission neutrons

For each fissile element, a different *pdf* has been defined for estimating the number of fission neutrons. (Note that, generally, these *pdf's* do not change significantly with the energy of the neutron inducing the fission process [4].)

Given $p(n)$ is the probability of the number of fission neutrons born from the fission event, then the FFMC for this discrete random variable n is obtained by satisfying the following inequality

$$P(n-1) < \eta \leq P(n), \quad for \ \ n = 0, n_{max} \tag{10.14}$$

where $P(n) = \sum_{n'=0}^{n} p(n')$.

In practice, however, rather than using Equation 10.14, the number of fission neutrons are sampled by using the average number of fission neutrons per fission ($\bar{\nu}$) given by

$$\bar{\nu} = \sum_{n'=0}^{n_{max}} n' p(n') \tag{10.15}$$

The procedure for sampling the number of fission neutrons using $\bar{\nu}$ is given below:

1. Generate a random number η

2. If $\eta \leq (\bar{\nu} - INT(\bar{\nu}))$, generate $INT(\bar{\nu}) + 1$ fission neutrons

3. If item 2 above is not true, then generate $INT(\bar{\nu})$ fission neutrons

Note that INT refers to a computer integer operation, which simply drops the digits beyond the decimal point.

10.3.1.2 Energy of fission neutrons

To sample the energy of fission neutrons, we sample from the fission neutron spectrum defined by

$$\chi(E)dE \equiv fraction\ of\ fission\ neutrons\ born\ within\ dE\ about\ E \tag{10.16}$$

Commonly, the fission spectrum is given by the *Watt* spectrum. For example, the *Watt* spectrum for *U-235* for thermal neutron fission is given by

$$\chi(\hat{E}) = 0.4527 e^{\frac{\hat{E}}{0.965}} sinh\left(\sqrt{2.29\hat{E}}\right) \tag{10.17}$$

where $\hat{E} = \frac{E}{E_0}$, and $E_0 = 1\ MeV$. An efficient methodology for sampling from this spectrum is presented in Chapter 2.

10.3.1.3 Direction of fission neutrons

Since the fission neutrons are emitted isotropically, their directions can be sampled from the following density function

$$p(\mu, \phi) = \frac{1}{4\pi},\ for\ -1 \leq \mu \leq 1\ and\ 0 \leq \phi \leq 2\pi \tag{10.18}$$

where μ refers to *cosine* of the *polar* angle, and ϕ refers to the *azimuthal* angle. The *pdf* for sampling μ and ϕ are derived below

$$p_1(\mu) = \frac{\int_0^{2\pi} d\phi p(\mu, \phi)}{\int_{-1}^1 d\mu \int_0^{2\pi} d\phi p(\mu, \phi)} = \frac{\frac{1}{2}}{1} = \frac{1}{2} \tag{10.19}$$

and

$$p_2(\phi) = \frac{p(\mu, \phi)}{\int_0^{2\pi} d\phi p(\mu, \phi)} = \frac{\frac{1}{4\pi}}{\frac{1}{2}} = \frac{1}{2\pi} \tag{10.20}$$

And the corresponding FFMC's are given by

$$\mu = 2\eta - 1 \tag{10.21}$$

and

$$\phi = 2\pi\eta \tag{10.22}$$

10.3.2 Procedure for Monte Carlo Eigenvalue simulation

As discussed earlier, in a criticality (eigenvalue) problem, the fission source, i.e., its spatial distribution is not known; therefore, in order to solve an eigenvalue problem, one has to assume/guess an initial fission source distribution and eigenvalue.

The straightforward *power-iteration* procedure for performing a Monte Carlo criticality calculation is described in 10.1.

There are a few difficulties (or flaws) with the *straightforward* procedure (Table 10.1). These are:

Table 10.1 *Straightforward* procedure for an eigenvalue Monte Carlo calculation

1	Partition the region containing fissile materials (*fuel*) into I subregions
2	Distribute N_0 fission neutrons over the subregions, e.g., the density of fission source per subregion, F_i, is set to $\frac{N_0}{I}$
3	For each fission neutron, sample energy and direction of the neutron
4	Transport each fission neutron through the model, until the neutron is absorbed or escaped.
5	If a neutron is absorbed in the fuel, then sample number of fissions, number of *new* fission neutrons, and their directions and energies. These *new* fission neutrons represent the next generation (or *cycle*) fission neutron source.
6	Repeat Steps 4 and 5 for n cycles, until the following convergence criteria are satisfied:

$$max\left|\frac{F_i^{(n)}-F_i^{(n-1)}}{F_i^{(n-1)}}\right| < \varepsilon_1, \text{ and}$$

$$\left|\frac{K^n-K^{n-1}}{K^{n-1}}\right| < \varepsilon_2$$

where, $K^{(n)} = \frac{\sum_{i=1}^{I} F_{i=1}^{(n)}}{\sum_{i=1}^{I} F_i^{(n-1)}}$

Note Typical values for ϵ_1 and ϵ_2 are in ranges of $10^{-2} - 10^{-4}$, and $10^{-4} - 10^{-6}$, respectively.

a. The fission neutron source $(F_i^{(n)})$ may not have converged, so one has to *skip* a number of cycles, i.e., n_s, which has to be tested.

b. The relative change in K from one cycle to the next might be very small and therefore masked by the statistical uncertainty. To overcome this difficulty, we evaluate a cumulative average value, k_c, given by

$$K_c^{(n)} = \frac{1}{n - n_s} \sum_{n'=n_s+1}^{n} K^{(n')} \tag{10.23}$$

c. If the expected eigenvalue (K) of the system of interest is not equal to 1, i.e., *subcritical* $(K < 1)$ or *supercritical* $(K > 1)$, then using the *straightforward* procedure (Table 10.1) will lead to an exponential drop or increase of neutron population (i.e., $N^{(n)}$). This is demonstrated in the following two examples, which estimate the neutron population for each fission cycle using

$$N^{(n)} = K^n N^{(0)} \tag{10.24}$$

Example 10.1 - Supercritical system

Consider $k = 1.2$ and $N^{(0)} = 1000$, then the expected number of fission neutrons as a function of cycle can be calculated by Equation 10.24. Table 10.2 presents the neutron population up to the 30th cycle.

The difficulty with a supercritical system is that the number of particles increases significantly from one cycle to another; therefore, we may use a significant amount of computer time while the source distribution has not converged.

Table 10.2 Expected fission neutron population as a function number of fission cycles

Cycle Number	Number of Neutrons
1	1,200
10	6,192
20	38,338
30	237,376

Table 10.3 Expected neutron poplution as a function number of cycles

Cycle Number	Number of Neutrons
1	800
10	107
20	12
30	1

Example 10.2 - Subcritical system

Consider $k = 0.8$ and $N^{(0)} = 1000$, then the expected number of fission neutrons as a function of cycle can be calculated by Equation 10.24. Table 10.3 presents neutron population up to 30 cycles.

For a subcritical system, the difficulty is that we may run out of particles before converging to a solution!

To address this issue, after every cycle, the fission neutron source is normalized as

$$q_i^{(n)} = \left(\frac{F_i^{(n)}}{\sum_{i=1}^{I} F_i^{(n)}} \right) N^{(0)} \qquad (10.25)$$

This means that the total number of fission neutrons in the system (i.e., neutron population) remains constant.

Table 10.4 presents the *standard* procedure used for performing an eigenvalue Monte Carlo simulation. This *standard* procedure effectively addresses the shortcomings of the *straightforward* procedure (items a-c above).

Note that the reliability of the results has to be examined as discussed in Chapter 6 for a fixed-source Monte Carlo calculation.

10.3.2.1 Estimators for sampling fission neutrons

In order to improve the degree of confidence in an eigenvalue Monte Carlo simulation results, commonly fission neutrons are sampled using three different estimators as follows:

Collision estimator - At every collision site, the number of fission neutrons is estimated according to

$$F = w \frac{\sum_k \overline{\nu}_k f_k \Sigma_{f,k}}{\sum_k f_k \Sigma_{t,k}} \qquad (10.26)$$

Table 10.4 Standard procedure for an eigenvalue Monte Carlo simulation

1	Partition the region containing fissile materials ($fuel$) into I subregions.
2	Set the *eigenvalue parameters*: N_p, number of particles per cycle N_s, number of skipped cycles N_a, number of active cycles
3	Distribute N_p fission neutrons over subregions, e.g., the fission source density per subregion, F_i, is set to $\frac{N_p}{I}$
4	For each fission neutron, sample energy and direction of the neutron.
5	Transport each fission neutron through the model, until the neutron is absorbed or escaped.
6	If a neutron is absorbed in the fuel, then sample number of fissions, number of *new* fission neutrons (i.e., next cycle source).
7	Calculate the normalized source ($q_i^{(n)}$) using Equation 10.25 and its associated uncertainty.
8	If $n \leq N_s$, then repeat steps 4-7, otherwise, go to step 9.
9	Calculate the cumulative average eigenvalue, k_c, using Equation 10.23 and its associated uncertainty, and calculate any requested tallies and associated uncertainties and FOMs.
10	if $n \leq N_a$, repeat steps 4-9, otherwise end the simulation.

where w refers to a particle statistical weight, k refers to the kth fissile isotope, f_k refers to the atomic fraction of the kth isotope, and $\overline{\nu}_k$, $\Sigma_{f,k}$, and $\Sigma_{t,k}$ refer to number of fission neutrons per fission, fission cross-section, and total cross-section of the kth fissile isotope, respectively.

Absorption estimator - If the neutron is absorbed in the fuel, the number of fission neutrons is estimated according to

$$F = w \frac{\sum_k \overline{\nu}_k f_k \Sigma_{f,k}}{\sum_k f_k \Sigma_{a,k}} \tag{10.27}$$

where $\Sigma_{a,k}$ refers to the absorption cross-section of the kth fissile isotope.

Path-length estimator - As the neutron traces a distance d in the fuel region, the number of fission neutrons is estimated according to

$$F = w \cdot d \sum_k \overline{\nu}_k f_k \Sigma_{f,k} \tag{10.28}$$

Note that each estimator can be more effective (achieving accurate and precise results) in a specific physical condition. For example, in a low-density medium, the path-length estimator is very effective, while in a dense medium with high absorption probability, e.g., mixture of fuel and moderator, the collision estimator is very effective. The simultaneous use of the three different estimators, however, increases the confidence on the expected source distribution and eigenvalue. In the following section, we present a methodology to combine the results of the three estimators.

10.3.3   A method to combine the estimators

To estimate a combined eigenvalue, Urbatsch et al. [102] present the necessary formulations that have been implemented into the Monte Carlo N-particle (MCNP) code system. This combined eigenvalue is obtained by using the least squares method, taking into account variances and co-variances of the three estimators, based on a paper by Halperin [47]. Formulations for both two- and three-estimator cases are derived and examined for a few problems. It is concluded that the three-estimator case is *almost* an optimum estimator.

Formulations for the three-estimator eigenvalue and its variance are presented below. For simplicity, if we represent k_{eff} by a variable (x), then the formulation for the three-estimator eigenvalue is given by

$$x = \frac{\sum_{\ell=1}^{3} f_\ell x_\ell}{\sum_{\ell=1}^{3} f_\ell} \tag{10.29}$$

where $x_\ell = k_{eff,\ell}$, ℓ refers to the estimator type, e.g., *collision*, *absorption*, or *path − length*, and f_ℓ is given by

$$f_\ell = S_{jj}^2(S_{kk}^2 - S_{ik}^2) - S_{kk}^2 S_{ij}^2 + S_{jk}^2(S_{ij}^2 + S_{ik}^2 - S_{jk}^2) \tag{10.30}$$

where each estimator (ℓ) uses a specific partial permutation of i, j, and k as listed in Table 10.5. Here, variances and covariances are expressed by

$$S_{ii}^2 = \frac{1}{N-1} \sum_{m=1}^{N} (x_{im} - \overline{x}_i)^2,$$ (10.31)

and

$$S_{ij}^2 = \frac{1}{N-1} \sum_{m=1}^{N} (x_{im} - \overline{x}_i)(x_{jm} - \overline{x}_j),$$ (10.32)

where N is number of cycles, and the summation in the denominator of Equation 10.29 is given by

$$f_{sum} = \sum_{\ell=1}^{3} f_\ell = S_{11}^2 S_{22}^2 + S_{11}^2 S_{33}^2 + S_{22}^2 S_{33}^2 +$$
$$2(S_{12}^2 S_{13}^2 + S_{22}^2 S_{13}^2 + S_{33}^2 S_{12}^2) -$$
$$2(S_{11}^2 S_{23}^2 + S_{22}^2 S_{13}^2 + S_{12}^2 S_{33}^2) -$$
$$(S_{12}^2 + S_{13}^2 + S_{23}^2)$$ (10.33)

Table 10.5 Partial permutations

ℓ	i	j	k
1	1	2	3
2	2	3	1
3	3	1	2

The combined variance associated with the three k_{eff} estimators is given by

$$S_{k_{eff}}^2 = \frac{S_1}{N \times f_{sum}} \left[1 + N \left(\frac{S_2 - 2S_3}{(N-1)^2 f_{sum}} \right) \right]$$ (10.34)

where

$S_1 = \sum_{\ell=1}^{3} f_\ell S_{1\ell}^2$.

$S_2 = \sum_{\ell=1}^{3} (S_{jj}^2 - S_{kk}^2 - 2S_{jk}^2)\overline{x}_\ell^2$,

$S_3 = \sum_{\ell=1}^{3} (S_{kk}^2 + S_{ij}^2 - S_{jk}^2 - S_{ik}^2)\overline{x}_\ell \overline{x}_j$.

For further details on the theory and derivation of the combined uncertainty for k_{eff}, the reader should consult Urbatsch et al. [102] and Halperin [47].

10.4 ISSUES ASSOCIATED WITH THE STANDARD EIGENVALUE MONTE CARLO SIMULATION PROCEDURE

The standard procedure (Table 10.4) of an eigenvalue Monte Carlo simulation requires a set of *eigenvalue parameters* including, N_p (number of particles per cycle), N_s (number of skipped cycles), and N_a (number of active cycles). Results of an eigenvalue simulation are highly dependent on an *appropriate* combination of these parameters.

Selection of an *appropriate* set of eigenvalue parameters is not straightforward, and is affected by the problem physics and user's requirements and needs. For example, if N_p is *small*, particle *undersampling* can occur that may result in *biased* results, while if N_p is *large*, because of high computation time per cycle, a limited number of active cycles can be performed and therefore limited precision and/or detail can be achieved. If N_s is small, the lack of source convergence will lead to inaccurate results. And the value of N_a is affected by N_s and degree of precision and detailed needed.

As discussed in Section 10.2, the *power-iteration* technique has difficulties in achieving accurate solutions for problems with HDR or loosely coupled regions. It is almost impossible to achieve an accurate (unbiased) solution for such problems in a reasonable amount of time. Additionally, the *power-iteration* technique is plagued by the cycle-to-cycle correlation that may result in biased and underestimated uncertainties [27].

To avoid achieving biased and/or imprecise results, a user has to consider the following measures:

- Use of diagnostic tests to examine the convergence of the fission source.

- Performance of a set of experiments for different combinations of the eigenvalue parameters.

- Evaluation of the source distribution in addition to the diagnostic tests.

In summary, in an eigenvalue Monte Carlo simulation, even if the estimated k_{eff} and the fission neutron source distribution (q_i) seem to be highly precise, both their values and associated uncertainties can be biased, because of the following facts:

1. The fission neutron source has not converged then the estimated k_{eff} is not reliable.

2. The cycle-to-cycle correlation (present in a *power-iteration* technique) can lead to a biased (underestimated) uncertainty.

3. Undersampling can lead to a biased k_{eff} and source distribution.

In the remainder of this chapter, we will address the above issues.

10.5 DIAGNOSTIC TESTS FOR SOURCE CONVERGENCE

Here, we will introduce two source convergence diagnostic techniques: i) Shannon entropy; and, ii) center-of-mass (COM).

10.5.1 Shannon entropy technique

This is a commonly used technique in widely used Monte Carlo codes such as MCNP and Serpent. Hence, its concept and relevance are introduced and discussed and a detailed supporting derivation is provided in Appendix B.

10.5.1.1 *Concept of Shannon entropy*

To introduce the concept of the Shannon entropy, it is necessary to introduce the concept of a time series [54]. A time series is a sequence of successive points in time spaced at uniform time intervals. Observation of k_{eff} at different generations can be considered as a time series. One of the most important decisions (often assumptions) made is whether the time series is said to be stationary. If stationary, the stochastic process can be described by its mean, variance, covariance, and other related parameters. This is important, because a diagnosis of stationarity implies that the statistical properties of the series do not change with time (or cycle, in this case). Physically, without a stationary series, the source has not converged and reliable estimates of random variables, such as flux, k_{eff}, neutron current, etc., are not achievable. Note that even with a diagnosis of stationarity, bias can still be present in both the tallies and their uncertainties. This is addressed in the next section.

General statistical tests for stationarity are difficult to find. The major issue is the amount of information needed to effectively track the source distribution from one cycle to the next. One approach is to use the Shannon entropy [90]. From information theory, the entropy is considered as a measure of the minimum number of bits for representing a probability density on a computer, or a measure of information needed for predicting the outcomes of an experiment. In order to effectively demonstrate why Shannon entropy is a good parameter for detection of convergence of fission source distribution, we will present the Shannon entropy formulation based on [93, 21].

Consider an experiment of m possible outcomes with probability p_i for each outcome. Then, the corresponding Shannon entropy (H) is given by

$$H = S(p_1, p_2, p_3, \ldots\ldots p_m) = -C \sum_{i=1}^{m} p_i \log_2 p_i \qquad (10.35)$$

where c is an arbitrary constant. A detailed derivation of the above formulation is given in Appendix 6.

10.5.1.2   Application of the Shannon entropy to the fission neutron source

We can directly apply the Shannon entropy to an eigenvalue Monte Carlo calculation. If we substitute the normalized fission source for each subregion, i.e., q_i in Equation 10.25 as p_i, and set the constant coefficient 1, then the entropy formulation for the fission neutron density is expressed by

$$H_s^{(n)} = -\sum_{i=1}^{m} q_i^{(n)} \log_2 q_i^{(n)} \qquad (10.36)$$

Where n refers to the cycle number.

One way to use Equation 10.36 is to examine the behavior of H from one cycle to next; if the source has converged, then it is expected that H fluctuates about an average value. However, the summation in Equation 10.36 may result in compensation of terms, which can lead to a false converge for high dominance ratio or loosely coupled problems in which the source distribution does not change much; therefore, H remains relatively constant for the same unconverged source distribution. Ueki and Brown [100, 101, 99] discuss alternative entropy formulations using other convergence criteria.

10.5.2   Center of Mass (COM) technique

To overcome the shortcoming of the Shannon entropy, Wenner and Haghighat [119] developed the source center-of-mass (COM) technique, which uses the COM of the source distribution to track the behavior of the source from one cycle to the next. If we consider vector position

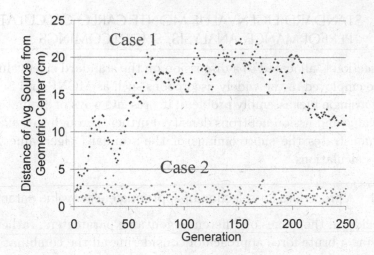

Figure 10.1 Distance of the source from the geometric center for Case 1 (vacuum boundary condition) and Case 2 (reflective boundary condition)

of each source subregion relative to the geometric center of the model as $\overrightarrow{r_i}$, then the vector position of the source COM is given by

$$\overrightarrow{R}^{(n)} = \sum_{i=1}^{N} q_i^{(n)} \overrightarrow{r_i} \tag{10.37}$$

where n refers to the cycle number, and $\overrightarrow{R}^{(n)}$ refers to the source COM for n^{th} cycle.

For example, in Wenner and Haghighat [119, 117] and Wenner [116], a loosely coupled problem with a symmetric source distribution is examined with two different boundary conditions (BCs): i) *reflective* BC (Case 1); ii) *vacuum* BC (Case 2). Figure 10.1 shows the position of the source COM as a function of generation (or cycle) number for the two cases.

The symmetry condition requires that the COM fluctuates about the geometric center of the model. This is observed for Case 1 which employs a *reflective* BC, but not for Case 2. The reason for lack of convergence of Case 2 is that the *vacuum* BC results in the loss of boundary particles, causing significant undersampling, which cannot be resolved with the standard eigenvalue procedure (Table 10.4). Note that the Shannon entropy technique cannot identify this problem. For further discussions on this problem and the development of the stationarity test for the COM, the readers should consult [118].

10.6 STANDARD EIGENVALUE MONTE CARLO CALCULATION - PERFORMANCE, ANALYSIS, SHORTCOMINGS

This section is allocated to a discussion on the standard eigenvalue procedure employed in the widely used codes such as MCNP and Serpent. Using a single fuel assembly problem, it presents ways of evaluating the eigenvalue and fission neutrons density. Further, through this example, it demonstrates the shortcomings of the standard eigenvalue Monte Carlo calculations.

10.6.1 A procedure for selection of appropriate eigenvalue parameters

To decide on the values of different eigenvalue parameters, rather than following a brute force approach of considering all the combinations of the three parameters, the following procedure is recommended:

1. Set the number of particles per generation (N_p) such that there are sufficient number of particles per tally region; this is especially important for the peripheral regions of models that can be affected by undersampling.

2. For a given N_p, examine the effect of number of skipped generations (N_s) for a modest number of active generations (N_a). Here, one may use the Shannon entropy and/or COM techniques to examine the source convergence.

3. After deciding on the value of N_s, examine the effect of increasing number of N_a by comparing *norms* of *relative differences* of k_{eff}, fission neutron source distributions, and associated statistical uncertainties.

4. After setting the N_a and N_s, examine increasing or decreasing number of N_p by repeating steps 2 and 3 above.

10.6.2 Demonstration of the shortcomings of the standard eigenvalue Monte Carlo calculation

Here, we try to demonstrate the shortcomings of a Monte Carlo eigenvalue calculation using the *standard* procedure (Table 10.4). For this, we will use an example from the recent work by Mascolino et al. [71] who address the shortcomings of an eigenvalue Monte Carlo calculation by performing a detailed study for a spent fuel assembly and a full

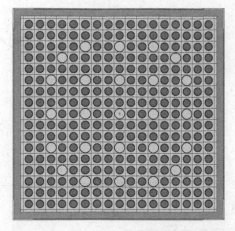

Figure 10.2 Standard 17×17 fuel assembly, surrounded by the cask material and absorber plates in between the adjacent assemblies.

spent-fuel cask with 32 assemblies, i.e., the GBC-32 cask benchmark problem. [105]

10.6.2.1 Example problem

The selected example is a single spent fuel assembly [71], which is a standard fuel assembly (with a 17×17 fuel lattice), flooded with water and surrounded by absorber plates, as shown in Figure 10.2. Further, the problem boundaries are represented by using the reflective boundary condition along the x and y boundaries, and the vacuum boundary condition along the z boundaries. It is worth noting that the presence of the absorber plates creates a unique difficulty that is the potential for undersampling near the lattice boundaries.

10.6.2.2 Results and analysis

To be able to perform a detailed analysis, the fuel region with *264 fuel rods* is partitioned into 24 axial segments, resulting in 6336 tally regions (I). Because of the small size of each tally region, we consider a large number of particles per cycle, i.e. N_p is set to 10^6. This results in about 158 particles per tally region. Further, we set a large number of skipped cycles (N_s) of 1000, and examine the behavior of eigenvalues and fission densities for the increasing number of active cycles. To examine the source convergence, we employ the Shannon entropy diagnostic test.

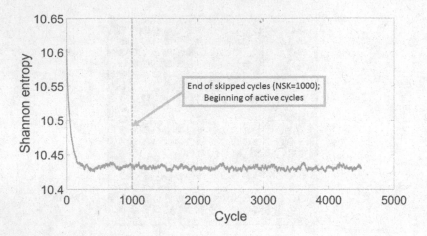

Figure 10.3 Behavior of the Shannon entropy

Figure 10.3 shows the Shannon entropy for the fission source distributed over the 6336 subregions for the increasing number of the active cycles (N_a) in range a of 100 to 3500. These results demonstrate that the Shannon entropy has settled beyond about 200 skipped cycles, and therefore one may conclude that the source has converged. This is explored further by examining the k_{eff} and the fission source distribution.

Figures 10.4 and 10.5 show the behavior of the k_{eff} and its relative uncertainty for the increasing number of active cycles. Again, these figures indicate that the estimated eigenvalue follows the expected behavior, i.e., k_{eff} is settled and the associated uncertainty drops by $\frac{1}{\sqrt{N_a}}$. These results indicate that the estimated k_{eff} are precise as they follow the Central limit theorem.

To examine the precision of the source distribution, we determine the relative difference of the source distribution for a given N_a to that of a reference case with N_r cycles, as follows

$$rd_i^{(N_a)} = \frac{q_i^{(N_a)} - q_i^{(N_r)}}{q_i^{(N_r)}}, \quad for \ N_a < N_r \tag{10.38}$$

where $rd_i^{(N_a)}$ refers to the source relative difference for i^{th} fission region for a case with N_a active cycles, $q_i^{(N_a)}$ refers to the fission source density at i^{th} fission region for the case with N_a active cycles, and $q_i^{(N_r)}$ refers the fission source density for the reference case with N_r active cycles. For this study, N_r is set to 3500 cycles. Then, we determine the L_1

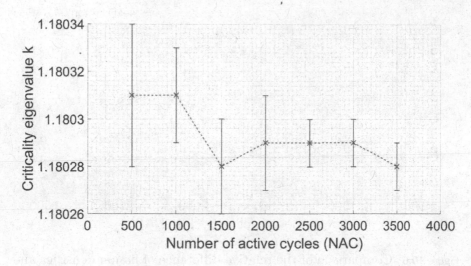

Figure 10.4 Behavior of k_{eff}

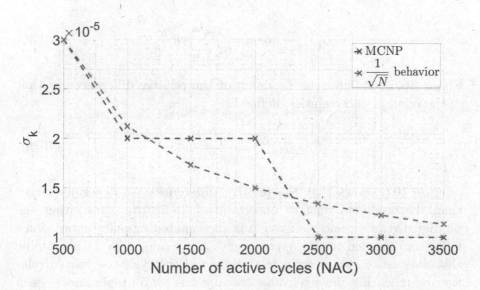

Figure 10.5 Behavior of k_{eff} uncertainty (Note that the MCNP results are provided with 5 significant figures, only)

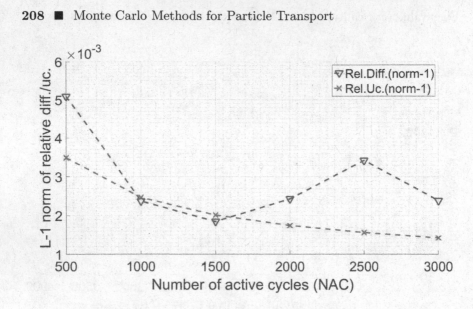

Figure 10.6 Comparison of the relative difference of fission densities and their associate statistical uncertainties

norm of the relative differences and compare it to the code-predicted statistical uncertainties. The formulation of the L_1 *norm* is given by

$$L_1^{(N_a)} - norm = \frac{1}{I} \sum_{i=1}^{I} rd_i^{(N_a)} \tag{10.39}$$

Figure 10.6 compares the L_1 *norm* of the relative differences to that of the relative uncertainties, defined by

$$R_i^{(N_a)} = \frac{S_i^{(N_a)}}{q_i^{(N_a)}}. \tag{10.40}$$

Figure 10.6 shows that the relative differences are generally larger than the predicted relative uncertainties, indicating that either the predicted relative uncertainty is underestimated or many more simulations are needed for the increasing N_a, N_s, and/or N_p. To examine this observation, we investigate the effect of the cycle-to-cycle correlation by repeating the eigenvalue calculations for 50 replications; each replication uses a different *seed* for the random number generator. To manage the computer time, we reduce the number of skipped cycles to 300, and the number of active cycles to 500, while using the same number of particles per cycle (i.e., $N_p = 10^6$).

The results of the replications are analyzed by comparing the average relative uncertainty predicted by the *standard power-iteration* algorithm to the *real* uncertainty obtained by performing 50 replications. The formulation of the average relative uncertainty for the standard algorithm is given by:

$$S_{code} = \frac{1}{N_{rep}} \sum_{n_r=1}^{N_{rep}} S_{n_r} \qquad (10.41)$$

And, the formulation for determination of the *real* relative uncertainty of fission densities is given by

$$S_{real} = \sqrt{\frac{1}{N_{rep}-1} \sum_{n_r=1}^{N_{rep}} (x_{n_r} - \overline{x_r})^2} \qquad (10.42)$$

where $\overline{x_r}$ refers to the average tally value over all replications. To compare the uncertainties, we determine the distribution of the ratios of the uncertainties given by:

$$f_s = \frac{S_{real}}{S_{code}} \qquad (10.43)$$

Figure 10.7 shows that the ratios (f_s) for the most part are larger than 2, indicating the significant underestimation of of the code-predicted uncertainties that is one of the major shortcomings of the *standard* algorithm. Table 10.6 compares the average values of the two uncertainties. These results indicate that the *code-predicted* uncertainty is

Table 10.6 Comparison of real and MCNP average uncertainties in fission source distribution

S_{real}	S_{MCNP}	f_s
0.72	0.35	2.28

underestimated by an average factor (f_s) of 2.28. Now, if we multiply the code-predicted uncertainties by the f_s values, we obtain an adjusted distribution of uncertainties.

Figure 10.8 clearly confirms that the estimated relative differences of source distributions for different number of cycles are within the range of *code-predicted* uncertainties and the *real* uncertainties. This

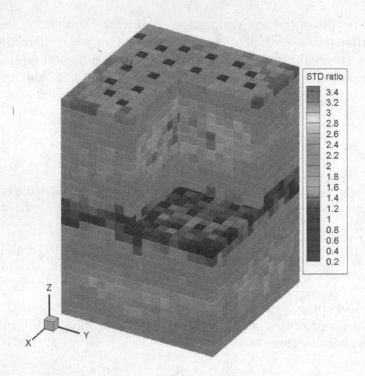

Figure 10.7 Distribution of the f_s for fission subregions in the single-assembly problem

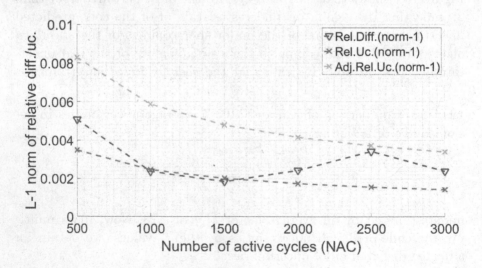

Figure 10.8 Comparison of the relative differences to the *real* and *code* uncertainties

means that the tallies are precise for a given number of active cycles, and the cycle-to-cycle correlation has indeed resulted in underestimating the uncertainties.

The discussions in this section clearly demonstrate the difficulties of the standard eigenvalue Monte Carlo calculations, and provide ways to analyze and overcome some of the difficulties. This, however, may not be practical for the most part for real-world problems because of the need for significant experience and computer resources. Hence, the author proposes alternative eigenvalue Monte Carlo techniques that are the subject of the next chapter.

10.7 REMARKS

This chapter introduces the eigenvalue Monte Carlo methodology for particle transport. It systematically develops a standard procedure commonly used in widely used Monte Carlo codes. It elaborates on different estimators and how the results can be effectively combined. The chapter highlights the difficulties of the standard Monte Carlo calculations, and possible techniques to identify them. Using a real-world problem, it demonstrates a process for evaluating the estimated k_{eff} and source distribution and their uncertainties. It also demonstrates ways to address the shortcomings of the standard procedure. However, it concludes with recommendation for using alternative techniques that are discussed in the next chapter.

PROBLEMS

1. Consider that the probability density of the number of fission neutrons emitted per fission of U-235 is given in Table 10.7 [115] for two energies of the neutron inducing fission.

    a. Determine the average number of fission neutrons and associated uncertainties for the two neutron energies.

    b. Considering the average and variance obtained in part (a), use a normal distribution to determine the fraction of fission neutrons between 0 and 5 neutrons.

    c. Compare your results to those given in Table 10.7.

2. Write a program to sample the fission spectrum, Equation (10.17), using the numerical inversion technique. Compare the performance of your algorithm to that presented in Chapter 2.

Table 10.7 Number of neutrons emitted per fission

Neutrons per fission	Probability	
	E=80 keV	E=1.25 MeV
0	0.02	0.02
1	0.17	0.11
2	0.36	0.30
3	0.31	0.41
4	0.12	0.10
5	0.03	0.06

3. Using the fission spectrum as the spectrum of neutrons inducing the fission process, write a program to determine the average fission neutron emission density using the information given in Table 10.7; for $E \leq 80$ keV; use the spectrum given for 80 keV, and, for $E > 80$ keV, use the spectrum given for 1.25 MeV.

4. Write a program to determine the eigenvalue and fission neutron distribution in a one-region, homogeneous slab reactor that is placed in a vacuum. Here, use the collision estimator, Equation (10.23).

   a. Test your program using the parameters from Table 10.8.

   b. Compare the shape of the fission density to a cosine function by using the test.

   c. If the reactor is not critical, revise your program so that you can search for a critical size.

5. Modify your program in Problem 4 to sample fission neutrons using the absorption and free-flight estimators, Equations (10.27) and (10.28). Combine the three eigenvalues by using the formulation given by Equation 10.29. Compare your results to that of a simple average with equal weight.

Table 10.8 Slab reactor parameters for Problem 4

Σ_t (cm^{-1})	Σ_a (cm^{-1})	$\bar{\nu}\Sigma_f$ (cm^{-1})	$\bar{\nu}$	Size (cm)
3.7	0.6	0.7	2.43	88.83

6. Modify your program in Problem 4 to evaluate the generation-dependent Shannon entropy and center of mass.

   a. By plotting your results, examine how the two approaches provide information on the convergence behavior of the fission density.

   b. Repeat this calculation for a problem with 5 times the width. Note the changes in convergence rate.

CHAPTER 11

Fission matrix methods for eigenvalue Monte Carlo simulation

CONTENTS

11.1 INTRODUCTION

In the previous chapter, we elaborated on the *standard* eigenvalue Monte Carlo algorithm that is used in the majority of the publicly available Monte Carlo codes. We identified the shortcomings of the *standard* algorithm, and offered techniques for detecting false convergence and biased tallies and their associated statistical uncertainties. Considering the need for performing an extended set of analysis to determine the *appropriate* set of eigenvalue parameters, the need for unreasonable amount of computer resources and time for resolving issues such as HDR and undersampling, and the difficulty of addressing the inherent cycle-to-cycle correlation, the author recommends the use of alternate schemes, which are discussed in this chapter.

Hybrid deterministic-Monte Carlo methodologies offer techniques that for the most part avoid all the issues facing the *standard* Monte Carlo algorithm. Computer code systems such as COMET [84] and RAPID [45, 112] write the eigenvalue transport equation in form of a system of equations, and pre-calculate the expansion coefficients or matrix coefficients by performing a series of fixed-source Monte Carlo calculations.These methodologies have two major advantages: i) an eigenvalue problem is replaced with a series of fixed-source Monte Carlo calculations; ii) the system of equations can be solved in real time, as it uses the pre-calculated coefficients. Such methodologies are excellent candidates for performing parametric studies, as the coefficients can be calculated for different ranges of system parameters, and therefore a large number of analyses can be performed in parallel in a very short time. This is especially important for design of new systems and optimization and evaluation of operating systems.

In this Chapter, we focus on the RAPID methodology that uses the Fission-Matrix (FM) methodology. This methodology casts the linear Boltzmann equation into a matrix form, which can be solved via different techniques. The following sections elaborate on the FM methodology, techniques for solving the FM formulation, its implementation, and its accuracy and performance for simulating real-world nuclear systems

11.2 DERIVATION OF FORMULATION OF THE FISSION-MATRIX METHODOLOGY

The Fission-Matrix (FM) methodology is derived by rewriting the linear Boltzmann equation (LBE) in a matrix form. Starting with

Equation (10.1) given by

$$H\psi = \frac{1}{k}F\psi \qquad (11.1)$$

we rewrite the fission operator (F) as

$$F = \chi\tilde{F}$$

where,

$$\tilde{F} = \frac{1}{4\pi}\int_0^\infty dE' \int_{4\pi} d\Omega' \nu\Sigma_f(\overrightarrow{r}, E') \qquad (11.2)$$

Then, we multiply Equation 11.1 by H^{-1} to obtain

$$\psi = \frac{1}{k}H^{-1}\chi\tilde{F}\psi \qquad (11.3)$$

Now, if we multiply the above equation by $\tilde{F}$, we obtain the formulation for the FM methodology as follows:

$$\overrightarrow{q} = \frac{1}{k}\mathbf{A}\overrightarrow{q}$$

where

$$\overrightarrow{q} = \tilde{F}\psi, \qquad (11.4)$$

and

$$\mathbf{A} = FH^{-1}$$

Here, $\overrightarrow{q}$ refers to the fission neutron source vector with I tally regions, and $\mathbf{A}$ refers to the coefficient matrix.

In practice, the FM method has been implemented in two ways, including: (1) pre-calculation of elements of $\mathbf{A}$, and its use for determination of the system eigenvalue and eigenfunction; and, (2) on the fly calculation of elements of $(\mathbf{A})$, and its use for determination of the system eigenvalue and eigenfunction.

11.2.1 Implementation of the FM method - Approach 1

Equation 11.4 indicates that $\mathbf{A}$ operates on the fission source to obtain the next generation fission source as follows

$$\mathbf{A}q(\overrightarrow{p}) = \int_{\mathbf{p}'} d\mathbf{p}' a(\overrightarrow{p'} \to \overrightarrow{p})q(\overrightarrow{p'}) \qquad (11.5)$$

where $a(\overrightarrow{p'} \to \overrightarrow{p})$ gives the expected number of fission neutrons in phase space $d\mathbf{p}$ due to one fission neutron in phase space $d\mathbf{p}'$.

To utilize the above formulation, it is necessary to discretize the phase space, and replace the integration with a summation [118, 112]. For example, considering the spatial variable only, the fission source density in a discretized spatial mesh i is obtained by

$$q_i = \frac{1}{k} \sum_{j=1}^{I} a_{i,j} q_j, \qquad for \ i = 1, I \tag{11.6}$$

where q_i is the fission source in mesh volume i, I refers to the total number of spatial meshes, and elements $a_{i,j}'s$ are obtained by using the following formulation

$$a_{i,j} = \frac{\int_{v_i} d^3 r \int_{v_j} d^3 r' a(\overrightarrow{r'} \to \overrightarrow{r}) q(\overrightarrow{r'})}{\int_{v_j} d^3 r' q(\overrightarrow{r'})} \tag{11.7}$$

The $a_{i,j}$ coefficient refers to the expected number of fission neutrons born in mesh volume (v_i) due to a fission neutron born in mesh volume (v_j). Note that, in addition to the spatial variable, it is possible to discretize other variables of phase space including energy and angle and rewrite Equations 11.6 and 11.7 accordingly.

In this approach, first the $a_{i,j}$ coefficients are calculated by performing a series of fixed-source Monte Carlo simulations for neutron sources are placed in each mesh volume (v_i) and the resultant fission densities are determined in all mesh volumes $(j = 1, I)$ including mesh volume (i).

Here, however, we should not perform brute force calculations, i.e., $I \times I$ calculations; rather, we should avoid unnecessary calculations by considering problem physics such as localized coupling, model symmetries, and regional similarities. These are addressed in several papers by Haghighat et al. [45], Walters et al. [113, 48], Roskoff and Haghighat [86, 87], and Mascolino et al. [71, 74], and later in this chapter (Sections 11.3.1.2 & 11.3.3).

For completeness of this discussion, it is important to introduce the FM formulation for a subcritical system in which independent neutron sources cause subcritical multiplication in presence of fissile elements or isotopes. A good example for a subcritical system is a facility containing spent fuel assemblies. In such a system, radioactive materials in the spent fuel either directly (e.g., spontaneous fission) or indirectly (e.g., (α, n) interaction) generate neutrons; hereafter, they are referred to as *intrinsic source*. These sources in turn induce fission in the existing fissile elements; this process is referred to as *subcritical multiplication*.

To account for the *intrinsic sources*, the FM formulation is given by,

$$q_i = \sum_{j=1}^{I} (a_{i,j} q_j + b_{i,j} q_j^{in})$$ (11.8)

where q_j^{in} refers to the intrinsic sources, and $b_{i,j}$ refers to the elements of coefficient matrix ($\mathbf{B}$) for the intrinsic sources which are calculated using the same process as used for the fission neutrons.

In a subcritical system, in addition to k_{eff} (11.4) and source distribution, one should determine the subcritical multiplication factor (M) expressed by

$$M = \frac{\sum_i q_i + q_i^{in}}{\sum_i q_i^{in}}$$ (11.9)

This FM approach has been implemented into the RAPID code system, and its accuracy and performance demonstrated against reference Monte Carlo calculations.

11.2.2   Implementation of the FM method - Approach 2

In this approach, elements of the coefficient matrix ($\mathbf{A}$) are determined on the fly and used to determine the system eigenvalue and eigenfunction. This approach is referred to as Fission Matrix-Based Monte Carlo (FMBMC) ([28], [118]). Table 11.1 provides a procedure for the implementation of the FMBMC approach.

As Table 11.1 indicates, this approach relies on the use of the eigenvalue parameters, and therefore is affected by similar issues as the *standard* eigenvalue Monte Carlo suffers from.

11.2.2.1   *Issues associated with the FMBMC approach*

Studies on the FMBMC Dufek [28]; Wenner and Haghighat [118] have indicated that the FM method can be effective for solving eigenvalue problems. The accuracy of the method depends on the accuracy of the fission matrix coefficients, which can be affected by the selected subregions for source distribution and the eigenvalue parameters. It is demonstrated that if the subregions are "small," then the elements of the FM are not affected by the source distribution. However, the

Table 11.1 Standard procedure for an eigenvalue FMBMC approach

1	Partition the region containing fissile materials ($fuel$) into I subregions.
2	Set the *eigenvalue parameters*: N_p, number of particles per cycle N_s, number of skipped cycles N_a, number of active cycles
3	Distribute N_p fission neutrons over subregions, e.g., the fission source density per subregion, F_i, is set to $\frac{N_p}{I}$.
4	For each fission neutron, sample energy and direction of the neutron.
5	Transport each fission neutron through the model, until the neutron is absorbed or escaped.
6	If a neutron from region i is absorbed in the fuel region j, then count fission neutrons using $a_{i,j} = a_{i,j} + w\frac{\overline{\nu}\Sigma_f}{\Sigma_t}$.
7	If $n > N_s$, then calculate cumulative average of $a_{i,j}$ $a_{i,j}^{(n)} = \frac{1}{n-N_s} \sum_{k=N_s+1}^{n} a_{i,j}^{(k)}$.
8	Calculate normalized ($a_{i,j}$), given by $\hat{a}_{i,j}^{(n)} = \frac{a_{i,j}^{(n)}}{q_i^{(n)}}$.
9	Use Equation 11.4 to determine K_{eff} and fission source (q_i) for generation n.
10	If $n \leq N_a$, repeat steps 4-9, otherwise end the simulation.

use of "small" meshes may be a daunting task computationally and, moreover, there is a need to verify that the subregions are adequately "small."

Because there is no formulation for selecting appropriate fission source subregions, it is necessary to investigate if the estimated matrix elements are reliable. To do this, one has to examine if there is any correlation among elements from different generations and estimate the uncertainties associated with each element. Additionally, knowing the uncertainties in the coefficient elements, it is necessary to develop a methodology for propagation of errors for determination of uncertainties in the estimated fission source distribution and corresponding eigenvalue. Wenner [116] and Wenner and Haghighat [118] present methodologies for examination of randomness and determination of uncertainties, therefore, providing the user with information on the reliability of the estimated elements of the fission matrix.

11.3 APPLICATION OF THE FM METHOD - APPROACH 1

Approach 1 (Section 11.2.1) that relies on a series of fixed-source calculations is the preferred technique, because it does not rely on the power-iteration technique and its shortcomings.

This approach is implemented into the RAPID code system, and used for simulation of various computational and experimental real-world problems. [112, 86, 88, 71, 113, 74, 73, 48].

In this section, we will present the application of the FM method to a spent fuel cask and a reactor core. We will compare the accuracy and performance of RAPID (i.e., FM technique) to that of the Serpent code system.

11.3.1 Modeling spent fuel facilities

Spent fuel assemblies, commonly placed in pools and casks, are separated by plates that contain neutron absorbing materials such as Boral. To maintain the criticality safety of a spent fuel pool/cask, it is necessary to demonstrate that the system is in a subcritical condition.

A spent fuel facility may contain fuel assemblies of different burnups, cooling times, and/or initial enrichments. So, performing a best estimate calculation would require significant amount of engineers' time and computer resources. The RAPID code system with its FM algorithm (Section 11.2.1) offers an accurate and highly efficient way for simulation of a spent fuel facility. In this section, first, we will discuss the process of pre-calculation of the FM coefficients for a spent fuel facility, then we will compare the accuracy and efficiency of RAPID to that of a reference Serpent Monte Carlo calculation.

11.3.1.1 Problem description

We consider the GBC-32 cask benchmark [105] that was used in a recent benchmarking study by Mascolino et al. [74]. The model developed in this study, shown in Figure 11.1, includes 32 fuel assemblies placed in a canister with Boral plates (encased in Aluminum cladding) in between the fuel assemblies. Further, the canister is placed in a Stainless Steel cylinder and flooded with water. For this demonstration we

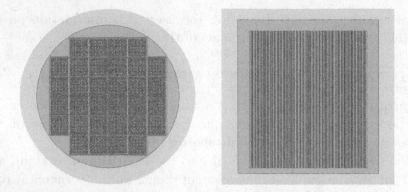

Figure 11.1 Serpent model of a spent fuel cask [radial (left) and axial (right) projections]

consider that all of the assemblies contain fresh fuel of 4% (weight) enrichment.

11.3.1.2 FM coefficient pre-calculation

A FM coefficients database is prepared as a function of different fuel burnup steps including 0 (fresh fuel), 10,000, 20,000, 30,000 and 40,000 MWd/MTHM using the the Serpent Monte Carlo code system. Since the fuel assemblies have an octal symmetry, coefficients are calculated for only 39 fuel rods/pins, shown in Figure 11.2, for the aforementioned six burnup steps. This means that a total of 234 fixed-source Serpent calculations are performed. In each of these calculations, the neutron source is placed in one 1-inch axial segment of each fuel pin and coefficients are tallied within each segment for 32" axially, and within the neighboring two assemblies away from the central assembly; this means that tallies are made in a 5x5 array of assemblies. Then using the symmetry and similarity conditions, the FM coefficients are set for the remainder of the model. For this demonstration database, the FM coefficients are calculated at 48 axial segments for each pin. For further detail on the determination of coefficients, see next sections, and consult Walters et al. [113] and Mascolino et al. [71].

Table 11.2 presents the total wall-clock for pre-calculations using the Serpent code system on 56 computer cores.

The above table indicates that with a modest number of computer cores of 56, we can prepare a database of FM coefficients as a function of burnup (fresh to 40,000 MWd/MT) in about 13 hours. Note that

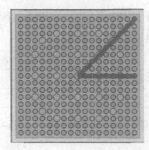

Figure 11.2 A fuel assembly (17x17 pins) with octal symmetry

Table 11.2 Computer time for one-time pre-calculation of FM coefficients

Number of calculations	Number of computer cores	Wall-clock time (min)
1	4	45
234	56	748

since the coefficients are independent of each other, they can be calculated in parallel, i.e., with 936 cores, all of the calculations can be completed in 45 min. Further, since a fixed-source Monte Carlo calculation is highly scalable, the wall-clock time for the pre-calculation can be reduced proportional to the increasing number of computer cores. Further discussion on the parallel processing of the Monte Carlo methods is provided in the next chapter.

11.3.1.3 Comparison of RAPID to Serpent - Accuracy and Performance

To measure the accuracy of the RAPID results, we performed a reference eigenvalue calculation using the Serpent code. For this calculation, we considered the following eigenvalue parameters including $N_p = 10^5$, $N_s = 250$, and $N_a = 500$. The fission neutron source is tallied in each pin in only 12 axial segments for achieving a reasonable precision in a reasonable amount of computation time.

Table 11.3 compares the k_{eff} predicted by RAPID to that of the reference Serpent calculation.

The pin-wise, axially dependent fission density distribution and its associated uncertainty predicted by Serpent are shown in Figures 11.3.

Table 11.3 Comparison of k_{eff}'s predicted by Serpent and RAPID for a full cask

Code	k_{eff}	Relative Difference (%)
Serpent	1.14546 ± 10 pcm*	-
RAPID	1.14570	1 pcm

* 1-σ statistical uncertainty in per cent mille (pcm), i.e., 10^{-5}

Figure 11.3 shows that the associated uncertainties are significantly larger than 10 % for the fission density at the top and bottom of the cask, and its periphery.

Figure 11.4 shows the fission neutron distribution using the RAPID code and the relative differences of the RAPID fission density distribution to that of the Serpent predictions.

The relative differences shown in Figure 11.4b indicate that the RAPID results are in an excellent agreement with the reference Serpent predictions. It is important to note that the large differences at the

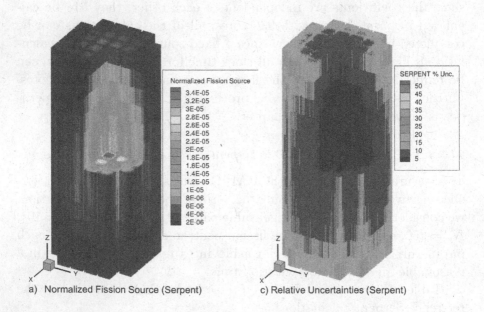

a) Normalized Fission Source (Serpent)   c) Relative Uncertainties (Serpent)

Figure 11.3 Fission neutron distribution and associated statistical uncertainty predicted by Serpent in a full cask

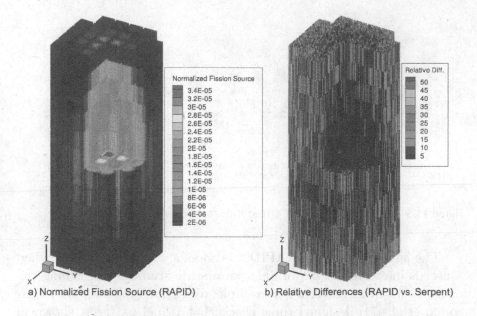

a) Normalized Fission Source (RAPID)   b) Relative Differences (RAPID vs. Serpent)

Figure 11.4 a) RAPID fission density distribution, b) Relative difference of the fission densities (RAPID versus Serpent)

top, bottom, and peripheral regions of the cask can be attributed to the large statistical uncertainties of the Serpent code.

To examine the performance of the two codes, in Table 11.4, we compare the wall-clock times for simulating a full spent fuel cask.

The above table demonstrates the significant advantage of the RAPID's FM matrix methodology over the Serpent's standard eigenvalue algorithm. It is important to note that to achieve acceptable uncertainties throughout the cask, the Serpent calculation would require significantly more particles per cycle (N_p) and more active cycles (N_a), i.e., requiring significantly more computation time.

Table 11.4 Performance of Serpent and RAPID - simulation of a full spent fuel cask

Code	Number of computer cores	Wall-clock time	speedup
Serpent	16	415 min	-
RAPID*	1	92 sec	271

* RAPID calculations performed in 48 axial segments

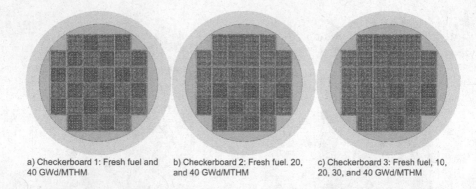

a) Checkerboard 1: Fresh fuel and 40 GWd/MTHM   b) Checkerboard 2: Fresh fuel. 20, and 40 GWd/MTHM   c) Checkerboard 3: Fresh fuel, 10, 20, 30, and 40 GWd/MTHM

Figure 11.5 Schematics of the three checkerboard cask cases

The advantage of the RAPID methodology is more evident when a user is interested in performing parametric studies or designing the most effective fuel pattern. For example, using the same database, Mascolino et al. [71] designed three checkerboard fuel patterns, shown in Figure 11.5, including: i) Case 1 - fresh and 40 GWd/MTHM fuels; ii) Case 2 -fresh and 20 and 40 GWd/MTHM fuels; and, iii) Case 3 -fresh and 10, 20, 30, and 40 GWd/MTHM fuels.

Table 11.5 compares the k_{eff}'s calculated by RAPID to that of the reference Serpent calculation for the three checkerboard cases.

The above table demonstrates that the RAPID's methodology yields accurate results even for casks containing assemblies of highly variable burnups. Table 11.6 compares the performance of the two codes for the three checkerboard cases.

Again, the above table demonstrates the significant speedups of the RAPID's FM matrix method. It is important to note that the Serpent results have significant uncertainties in the top, bottom and peripheral regions of the cask, in spite of considering only 12 axial segments versus the 48 axial segments used for the RAPID calculations.

Table 11.5 Comparison of K_{eff}'s for the checkerboard casks - RAPID versus Serpent

case	Serpent	RAPID	Relative difference (pcm*)
1	0.98679 ± 12 pcm	0.98693	14
2	1.03336 ± 11 pcm	1.03343	7
3	1.05311 ± 11 pcm	1.05336	25

* 1-σ statistical uncertainty in per cent mille (pcm), i.e., 10^{-5}

Table 11.6 Performance of the Serpent and RAPID - Checkerboard casks

Case	Serpent	Computer cores	RAPID*	Computer cores	Speedup
1	415 min	16	100 sec	1	249
2	421 min	16	100 sec	1	250
3	425 min	16	140 sec	1	182

* RAPID calculations performed in 48 axial segments

11.3.2 Reactor cores

In a spent fuel environment, because of the presence of absorbing plates, the range of impact of fuel assemblies is limited, and even the boundary wall or reflector has negligible effect on the peripheral assemblies. This situation, however, is quite different for a reactor core environment in which the range of influence of assemblies is longer, and boundary effect is significant. Additionally, a reactor core poses further complications because of the presence of the control rods and the use of burnable poisons. To address these issues, it necessary to develop novel techniques that effectively correct the FM coefficients, without the need for the generation of new FM coefficients. Recent work by Walters et al. [113] and He and Walters [48] offers innovative techniques to handle various conditions encountered in a nuclear reactor core.

Here, we present a few of the techniques [113, 48] that are incorporated into the RAPID code system, and tested based on the OECD/NEA Monte Carlo Performance benchmark [51], the Benchmark for evaluation and validation of reactor simulations (BEAVRS) [52], or their variations.

11.3.3 A few innovative techniques for generation or correction of FM coeffiicients

Here, we will introduce a few techniques for handling geometric similarity, boundary effects, and material discontinuity.

11.3.3.1 Geometric similarity

In most nuclear reactors or spent fuel systems, axial material distribution and geometry do not change rapidly. Hence, the FM coefficient from an axial position z to a distance d away $(z \rightarrow z + d)$ should be largely independent of the starting position z_1, and only dependent on the relative distance between the points, i.e., $(z_1 \rightarrow z_1 + d) \approx (z_2 \rightarrow z_2 + d)$. This means that FM becomes dependent on distance (d) only rather than both z_1 and z_2. Hence, fixed-source calculations need to be performed at only a single z-location, and the results can be translated up and down.

Additionally, in a nuclear reactor core there is a repeating assembly-wise geometry. For a core of one type of assembly, if a FM coefficient is known between two pins in one assembly, then a similar coefficient can be used for two other pins with similar relative positions in another assembly. This assumption, however, breaks down near the core boundaries. This means that calculations need to be performed only for the pins in one assembly, and the coefficients can be translated to other assemblies. Moreover, the octal symmetry of a standard PWR fuel assembly would reduce the number of calculations to the fuel pins in one octant of a fuel assembly, e.g. for a 17×17 lattice, FM coefficients have to be calculated for only 39 pins.

These measures siginficantly reduce the number of computations and amount of computer memory needed.

11.3.3.2   Boundary correction

To correct for axial and radial boundaries, a ratio correction method has been developed [111], and implemented into the RAPID code system. Assuming that the axial and radial correction factors are independent, then the boundary correction formulation is given by

$$\tilde{a}_{i,j} = a_{i,j} \times bnd(x,y) \times bnd(z) \qquad (11.10)$$

Where, $bnd(x,y)$ is the radial boundary correction factor given by

$$bnd(x,y) = \frac{F_{radial}(x,y)}{F_{infinite(x,y)}} \qquad (11.11)$$

And, $bnd(z)$ is the axial boundary correction factor given by

$$bnd(z) = \frac{F_{radial}(z)}{F_{infinite(z)}} \qquad (11.12)$$

Here, F refers to the fission neutron density at a tally region. In order to determine the boundary correction factors, four extra calculations have to be performed:

1. A **standard full core** eigenvalue calculation to obtain the fission density distribution (F) throughout the core; this fission density is used for the three models below.

2. An **inifinite** model without axial and radial boundaries using the fission density (F) as a fixed source to determine $F_{infinite(x,y)}$ and $F_{infinite(z)}$.

3. A **radial semi-infinite** model without axial boundaries using the fission density (F) as a fixed source to determine $F_{radial(x,y)}$.

4. An **axial semi-infinite** model without radial boundaries using the fissiond density (F) as a fixed source to determine $F_{axial(z)}$.

11.3.3.3 Material discontinuity

In a nuclear reactor, often material compositions or burnups of adjacent assemblies are different. Henceforth, the FM coefficients for the boundary pins are affected by the adjacent assemblies of different material compositions.

Here, we present one of the techniques developed by He and Williams [48]. Consider that the FM coefficients are known for two cases (1 and 2) with two assemblies of the same type as shown in Figure 11.6.

Now, let's make a mixed model by replacing the right-hand assembly of the model type 1 with a type 2 assembly as shown in Figure 11.8. Based on the methodology developed by He and Willaims [48], correction ratios for the mixed model are determined using the steps below:

1. Determine fission neutron density in the three models by performing fixed-source calculations using a uniform source distribution.

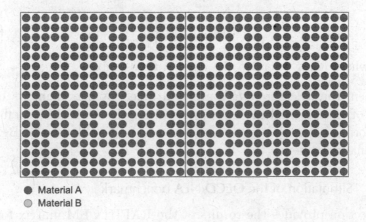

● Material A
○ Material B

Figure 11.6 Schematic of Case 1 - a two-assembly model of type 1

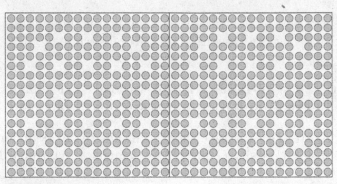

● Material A
○ Material B

Figure 11.7 Schematic of Case 2 - a two-assembly model of type 2

2. Determine pin-wise ratios defined by

$$R_i = \frac{F_i(real)}{F_i(uniform)} \qquad (11.13)$$

where the *uniform* case is chosen such that pin i is in the same type of assembly as the *real* case.

3. For example, for the current model, we have to determine the following:

$$R_i(left) = \frac{F_i(case\ 3)}{F_i(case\ 1)}, \qquad (11.14)$$

where *left* refers to the *left-side* assembly, and

$$R_i(right) = \frac{F_i(case\ 3)}{F_i(case\ 2)}, \qquad (11.15)$$

where *right* refers to the *right-side* assembly

Then, the calculated R_i's are multiplied by the elements of each row i of the original A matrix to obtain the corrected coefficients. This formulation is demonstrated successfully [48] for the BEAVRS benchmark problem.

11.3.4   Simulation of the OECD/NEA benchmark

This section provides the results of the RAPID's FM matrix formulation for simulation [113] of the OECD/NEA Monte Carlo Performance benchmark [51].

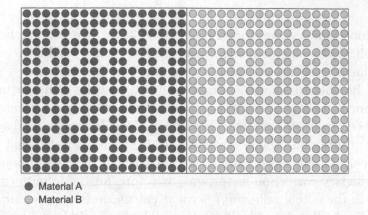

● Material A
◌ Material B

Figure 11.8 Schematic of Case 3 - a two-assembly model of mixed types 1 and 2

The benchmark represents a reactor core consisting of 241 identical fuel assemblies with a 17 × 17 lattice. The radial projection of the core is shown in Figure 11.9. Each fuel assembly contains 264 fuel rods and 25 guide tubes. Axially, in addition to the fuel active length of 366 cm, the bottom fuel assembly region, nozzle region, and core plate

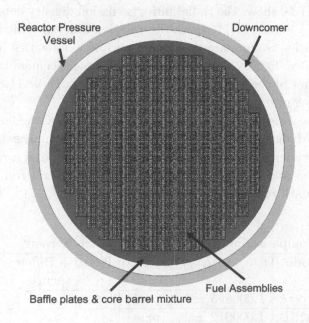

Reactor Pressure Vessel

Downcomer

Baffle plates & core barrel mixture

Fuel Assemblies

Figure 11.9 Radial projection of the OECD benchmark problem

region are modeled. Additionally, the benchmark considers: i) axially, two moderator temperatures, *cold* and *hot*, for the bottom half and top half of the reactor model, respectively; and, ii) radially, two fuel enrichment levels are considered, *low* and *high*.

To be able to simulate the benchmark with its temperature and enrichment variations, the FM coefficients are calculated for 3 assembly types (i.e., *cold*, *cold*, and *cold-enriched*). Considering an octal symmetry, i.e., containing 39 fuel pins, a total 117 fixed-source Serpent Monte Carlo calculations are performed with 3×10^7 in 105.3 CPU hours. For the boundary correction factor, only two (one for the infinite case and one with the actual reflectors) Serpent calculations were performed to determine $bnd(x, y)$ and $bnd(z)$ factors. These calculations required 599 CPU hours. This means that all the pre-calculations required a total time of about 700 CPU hours.

Here, we present the results for the uniform case without considering fuel enrichment and moderator temperature changes. Interested reader should consult [113] that examines various reactor conditions.

Table 11.7 compares k_{eff} determined by RAPID to that of the reference Serpent prediction.

The above results again demonstrate that the RAPID results in excellent agreement with the reference Serpent predictions.

Figure 11.10 shows the radial pin-wise fission density determined by RAPID, and Figure 11.11 shows the relative difference of the RAPID fission densities as compared to predicted fission densities by Serpent.

The above figures demonstrate excellent agreement of RAPID results to that of the Serpent predictions. Finally, Table 11.8 compares the computer resources used for both codes.

This study clearly highlights the significant advantage of using the RAPID's FM algorithm for performing reactor core analysis. Again, the interested reader is encouraged to consult [113, 48], and a recent work by Mascolino and Haghighat [73] that address control rod movement in a benchmark research reactor.

Table 11.7 Comparison of k_{eff}'s - RAPID versus Serpent

Code	K_{eff}	Relative Difference (pcm)
Serpent	1.000855 ± 1.0 (pcm)	-
RAPID	1.000912 ± 1.4 (pcm)	5.3

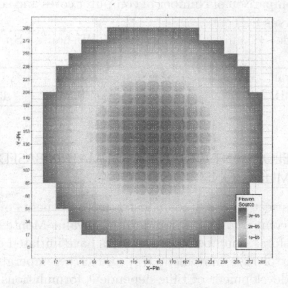

Figure 11.10 Radial projection pin-wise fission density calculated by RAPID

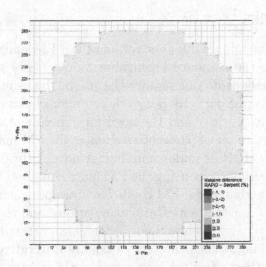

Figure 11.11 Relative differences (%) of the pin-wise fission densities - RAPID versus Serpent

Table 11.8 Comparison of Number of computer cores and computer time
- RAPID versus Serpent

Code	Number of computer cores	CPU-hours	Speedup
Serpent	20	1000	-
RAPID	1	0.23	4348

11.4 DEVELOPMENT OF OTHER FM MATRIX BASED FORMULATIONS

The previous section clearly demonstrates the benefits of the FM formulation as compared to the standard eigenvalue Monte Carlo formulation. Therefore, a number of researchers have initiated efforts on the
effect of control rods movement [73] and [97], temperature feedback
effects [85], development of time-dependent formulations for determination of 3-D fuel burnup [87] and simulation of reactor kinetics [61, 72].
These efforts can significantly improve reactor performance, efficiency
and safety, as they provide accurate solutions of high fidelity in relatively very short time.

11.5 REMARKS

This chapter introduces different forms of the FM matrix methods as
alternatives to the standard eigenvalue Monte Carlo technique. The
FM matrix methods do not require the use of the eigenvalue parameters (i.e., number of particles per cycle, number of skipped cycles, and
number of active cycles), and the need for experimentation for selecting an appropriate set of parameters. Hence, the FM matrix method is
not affected by HDR or undersampling. A novel hybrid deterministic-
Monte Carlo methodology using the FM matrix method has been developed. In this methodology, FM coefficients are pre-calculated through
a series of fixed-source Monte Carlo calculations, and the system eigenvalue (criticality) and eigenfuction (fission density distribution) are calculated by solving a linear system of equation iteratively. To use the
FM matrix method effectively for a real-world reactor problem, novel
correction techniques are developed for correcting the FM coefficients
to allow for simulation of cores loaded with variety of fuel designs,
without the need for performing more pre-calculations. The FM based

hybrid methodology has been incorporated into the RAPID code system and used for smulation of spent fuel facility and reactor cores. It is demonstrated that RAPID can successfully simulate spent fuel facilities and reactor cores in real time.

PROBLEMS

1. Consider a spent fuel pool containing two fuel assemblies and placed in vacuum as depicted in Figure 11.12. The width of fuel region is 15 cm and the width of water region is 2 cm. Considering the parameters given in Table 11.9.

   a. Simulate the problem using the program you developed in Problem 4, Chapter 10.

   b. Determine the fission matrix coefficients for each region.

   c. Develop a program based on Equation 11.6 to determine fission density and k eigenvalue.

   d. Compare your results from parts (a) and (c).

2. Modify your code from problem 10.4 to implement the standard fission matrix implementation (approach 2). Use an eigenvalue

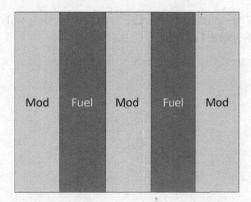

Figure 11.12   Slab reactor schematic for Problem 1

Table 11.9   Slab reactor parameters for Problem 1

Material	Σ_t (cm^{-1})	Σ_a (cm^{-1})	$\bar{\nu}\Sigma_f$ (cm^{-1})	$\bar{\nu}$	Size(cm)
Fuel	0.264	0.082	0.214	2.98	5 cm
Moderator	3.45	0.022	-	-	5 cm

solver (e.g., in Matlab) to obtain the fundamental eigenvalue and eigenvector. Compare the rate of convergence of the two approaches. (use the material and geometry data from problem 11.1)

3. Using the data from problem 11.2 in an eigenvalue solver, obtain the first two eigenvalues and calculate the spectral radius of the matrix (i.e., $\frac{k_1}{k_0}$). Double the size of the system and compare the resulting spectral radius.

4. Similar to 11.4, derive a formulation for determination of

   a. the multigroup fission density distribution,

   b. the total flux distribution,

   c. the multigroup flux distributions.

5. In a FM matrix methodology, one has to consider material and geometric symmetries and similarities for reducing the number of coefficients/calculations needed. One common condition in most nuclear reactors is the presence of geometric symmetries. For example, the fuel assemblies tend to have an octal symmetry. Determine the number of coefficients for

   a. a fuel assembly of 17x17 square lattice with 25 guide tube (see Figure 11.2).

   b. any assembly of arbitrary lattice size of $n \times n$.

CHAPTER 12

Vector and parallel processing of Monte Carlo particle transport

CONTENTS

12.1 INTRODUCTION

Since the early 1980s, there has been significant progress in computer hardware design. Besides designing faster computer chips, computer vendors have introduced vector and parallel hardware, which can increase the computer performance (GFLOPS:Giga FLoating point

OPerations per Second) by orders of magnitudes. The performance improvement due to vector or parallel processing is highly dependent on the algorithm, i.e., how well it suits and/or how well it is designed for vector/parallel architectures.

The trend for the design of faster computer chips has followed Moore's law [76], which was named after Intel co-founder Gordon E. Moore. Moore's law is the observation that, over the history of computing hardware, the number of transistors on integrated circuits doubles approximately every two years. However, it is predicted that this doubling period may slow down to three years in the near future.

One of the approaches for measuring parallel computer performance is to use the LINPACK software [26] for solving a system of 1,000 equations. *Summit* is a supercomputer developed by IBM for use at Oak Ridge National Laboratory; as of November 2019, *Summit* (with 164,000 cores) is the fastest supercomputer in the world, capable of 200 petaFLOPS. Its current LINPACK benchmark is clocked at 148.6 petaFLOPS.

Parallel and vector supercomputers have resulted in a significant increase in the computer performance, e.g., between 1995 and 2019 the performance has increased by a factor of over 870,000. This performance, however, can be realized if one designs new software, which can utilize the vector and parallel processing capabilities of the supercomputers. The need for new software for solving different engineering and scientific problems has resulted in a new multidisciplinary area called scientific computing, high performance computing, and/or parallel computing. In this chapter, we first introduce the concept of vector and parallel processing, and then we discuss different vector and parallel algorithms for Monte Carlo methods.

12.2 VECTOR PROCESSING

Vector processing refers to performing an operation on all or a group of elements of an array (or vector) simultaneously. This can be further explained if we examine a DO-LOOP as is processed on a conventional ("scalar") computer versus a vector computer.

12.2.0.1 Scalar computer

To explain the working of a *scalar* computer, let's consider a DO-LOOP as follows:

```
DO I = 1, 128
    C(I) = A(I) + B(I)
ENDDO
```

On a *scalar* computer, the interaction between computer memory and CPU for processing this DO-LOOP in a low-level (e.g., assembly) language is described in Table 12.1.

The above table indicates that for every index of the DO-LOOP, corresponding elements of A and B are transferred to registers and added together, then the result is transferred to another register and, subsequently, to a new memory location. The loop index is incremented by one and compared with its maximum value. If the condition is satisfied, another set of elements are processed; otherwise, the DO-LOOP is terminated. It is important to note that for every index of the DO-LOOP, computer cycles (time) are used for accomplishing the following operations: LOAD, ADD, STORE, and CHECK.

12.2.0.2   *Vector computer*

In contrast to a *scalar* computer, a *vector* computer transfers all (or groups) of the elements from memory to register, simultaneously. For example, a *vector* computer processes the aforementioned DO-LOOP as follows:

```
LOAD V1, A(1:128)
LOAD V2, B(1:128)
```

Table 12.1   Processing a DO-LOOP on a scalar computer

	Instruction	Description
	LOAD R1, 1	Load 1 to register R1
10	LOAD R2, A(R1)	Load A(R1) to register 2
	LOAD R3, B(R1)	Load B(R1) to register 3, and store it in register 4
	STORE C(R1), R4	Move the content of fourth register to the memory
	R1 = R1 + 1	Increase the index R1 of the do-loop
	JUMP 10 IF R1 $\leq$ 128	Go to 10 if R1 is $\leq$ 128

```
            V3=V1+V2
  STORE C(1:128), V3
```

Because vector operations work on all elements at once and there is no need for index checking, it is expected that the vector procedure will be significantly faster than the scalar approach. It is worth noting that there is some overhead associated with the initialization of the vector registers. Because of this overhead, if an array has less than four elements, then the scalar operation becomes more effective.

The only difficulty in vectorizing software is the fact that the elements being simultaneously vectorized must be *independent*. For example, the following DO-LOOP cannot be vectorized because of the interdependency of the elements of **A** array.

```
A(1) = 1.0
DO I=2, 128
    A(I)=A(I-1)+B(I)
ENDDO
```

Finally, vector computer vendors have developed new compilers with a vector option, which if invoked will vectorize all the "clean" DO-LOOPs. Also, these compilers generally offer "directives" allowing the user to overwrite the compiler's decisions.

12.2.1 Vector performance

The performance of a vectorized code is estimated on the basis of the code speedup defined by

$$speedup = \frac{CPU(scalar)}{CPU(vector)} \tag{12.1}$$

The theoretical speedup, or Amdahl's law [2] for vectorization, is given by

$$speedup(theoretical) = \frac{T_s}{(1 - f_v)T_s + f_v T_v} \tag{12.2}$$

where T_s refers to the CPU time for the scalar code, f_v refers to the vectorizable fraction of the code, and T_v refers to the CPU for the vector code. Now, if the numerator and the denominator of the above equation are divided by T_s, the formulation for the *theoretical speedup* reduces to

$$speedup(theoretical) = \frac{1}{(1 - f_v) + f_v \frac{T_v}{T_s}} \qquad (12.3)$$

If the ratio of T_v to T_s and f_v are known (or can be estimated), then one can estimate the effectiveness of the vector algorithm.

In order to achieve significant vector speedups, it is necessary to adjust the length of the DO-LOOPs according to the machine vector length. Different vector computers have different vector lengths; e.g., the first CRAY computer had a vector length of 64, while CRAY C90 has a vector length of 128.

12.3 PARALLEL PROCESSING

Parallel processing refers to being able to process different data and/or instructions on more than one CPU. This can be accomplished in two environments: (1) several computers that are connected via a network (local and/or remote), and (2) a parallel computer that is comprised of several CPUs. If parallel processing is performed on a network, it is referred to as *distributed computing*, while if it is done on a parallel computer it is referred to as *parallel computing*.

General libraries, such as the Parallel Virtual Machine (PVM) [37] and Message Passing Interface (MPI) [42] allow distributed computing on a group of computers and/or a distributed-memory parallel computer. Software, such as CRAY's Autotasking and IBM's Parallel FORTRAN (PF) are designed for performing parallel computing on a specific parallel computer.

Since the inception of computers, numerous architectures have been proposed and studied. These architectures can be classified into four groups based on the Flynn's taxonomy [34], which considers the number of concurrent instruction and data streams:

1. **SISD** (Single Instruction Single Data): A serial computer that does not have any parallelism in either instruction or data stream.

2. **SIMD** (Single Instruction Multiple Data): A parallel computer that processes multiple data streams through a single instruction.

3. **MISD** (Multiple Instruction Single Data): This taxonomy has not been considered!

4. **MIMD** (Multiple Instruction Multiple Data): A computer environment that allows multiple instruction on multiple streams of data.

Based on the SIMD and MIMD taxonomies, numerous parallel computer architectures have been designed and used for various applications. Between the two, MIMD is more flexible; therefore, the majority of parallel computers today are designed based on this taxonomy. There are two flavors of MIMD based architectures:

1. Shared-memory MIMD: Processors share the same memory or a group of memory modules. Examples are CRAY Y-MP, CRAY C90/J90, or IBM 3090/600.

2. Distributed-memory MIMD: Each processor has a local memory and information is exchanged over a network, i.e., message passing. Today's examples are Intel Paragon, IBM Blue Gene, Beowulf Clusters, and Cray.

To develop a parallel algorithm, one has to incorporate new parallelization instructions into the "standard" FORTRAN or C code. To achieve high performance from a parallel algorithm, in most cases it is necessary to completely restructure the old codes or to develop new algorithms.

12.3.1 Parallel performance

The performance of a parallel algorithm is measured based on the following factors:

1. *speedup*, defined by

$$speedup = \frac{wall\text{-}clock\ time\ (serial)}{wall\text{-}clock\ time\ (parallel)} \qquad (12.4)$$

2. *efficiency*, defined by

$$efficiency(\%) = 100 \times \frac{speedup}{P} \qquad (12.5)$$

where P refers to the number of processors used.

The above factors can be compared to the theoretical values, predicted by the Amdahl's law. The theoretical speedup for parallel processing is defined by

$$speedup(theoretical) = \frac{T_{sw}}{(1 - f_p)T_{sw} + f_p \left(\frac{T_{sw}}{P}\right)} \tag{12.6}$$

where T_{sw} is the wall-clock time for the serial code, and f_p is the parallelizable fraction of the code. Now if the numerator and the denominator of the above equation are divided by T_{sw}, the formulation for the theoretical parallel speedup reduces to:

$$speedup(theoretical) = \frac{1}{(1 - f_p) + f_p \left(\frac{1}{P}\right)} \tag{12.7}$$

Equation 12.7 is very useful because, as shown in Figures 12.1 and 12.2, it provides the upper limits for *speedup* and *efficiency*, respectively, for a given parallel fraction (f_p).

Further, the *theoretical speedup* can be used in conjunction with the *measured speedup* (Equation 12.4) to estimate the *parallel fraction*(f_v) of a code, thereby realizing the effectiveness of the parallel code for a given architecture.

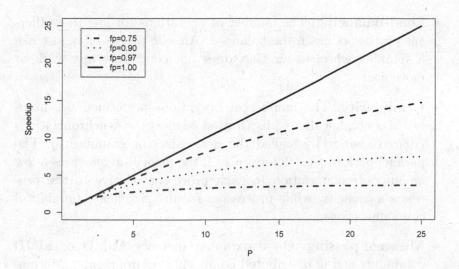

Figure 12.1 Demonstration of the Moore's law for *speedup*

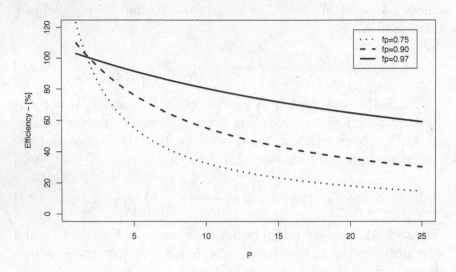

Figure 12.2 Demonstration of the Moore's law for *efficiency*

12.3.1.1 Factors affecting the parallel performance

Parallel performance of a parallel algorithm is affected by the following factors:

- **Load-balancing**: The number of operations (or load) on different processors has to be balanced. An *idle* processor is just like a spinning wheel in air that does not contribute any work or movement.

- **Granularity**: The number of operations performed per number of communications (distributed memory) or synchronizations (shared memory) is called the grain size (or granularity). The parallel performance deteriorates if the algorithm allocates a low amount of computation to each processor relative to the processor's capacity, while processors require significant number of communications.

- **Message passing**: On distributed memory MIMD or SIMD computers and in distributed computing environments, information is exchanged among processors over a network; this is called message passing. The number of messages and their size and relation to the network bandwidth affects the parallel performance.

- **Memory contention**: On shared memory MIMD computers, if different processors access (read/write) the same memory location, a memory contention would occur. This situation may result in several *idle* processors as they wait to access data.

12.4   VECTORIZATION OF THE MONTE CARLO PARTICLE TRANSPORT METHODS

The conventional *history-based* Monte Carlo cannot be vectorized because each history is made up of events/outcomes that are random. Therefore, different histories most likely require significantly different operations/calculations after the first step of simulation. Consequently, a vector of histories breaks down because its elements have to be processed through different arithmetic operations.

To overcome the obvious difficulty of the standard *history-based* algorithm, the alternative approach of *event-based* is considered. In the latter approach, histories are partitioned into a collection of events, such as free-flight, collision, boundary crossing, etc., which are similar and can be processed via a vector operation. This new approach is called an *event-based* algorithm.

The analog *event-based* Monte Carlo algorithm can be formulated in four events as listed below:

1. Event A: Free-flight

    a. Fetch atrributes of all particles.

    b. Perform free-flight sampling.

    c. Disperse attributes of all particles to one of the following events: collision; boundary crossing; or, termination.

2. Event B: Collision

    a. Fetch attributes of all particles.

    b. Perform collision sampling.

    c. Disperse attributes of all particles to one of the following events: termination or free-flight.

3. Event C: Boundary crossing

    a. Fetch attributes of all particles.

    b. Perform boundary crossing calculations.

    c. Disperse attributes of all particles to one of the following events: termination or free-flight.

4. Event D: Termination

There is no prescribed order in processing these events. Rather, at any time during the simulation, the event with the largest vector length is processed. To achieve significant vector speedups, it is necessary that the event vector length is comparable to the machine vector length.

The difficulty with the event-based formulation is the fact that a new software has to be developed; in other words, it is not practical to revise a *history-based* code to operate as an *event-based* algorithm. Brown and Martin [18] developed a new code and reported large speedups; e.g., 1–2 orders of magnitude for criticality calculations. One of the major difficulties of such as algorithm is development of clever bookkeeping logic, so that all the events are processed and accounted for in efficient manner.

12.5 PARALLELIZATION OF THE MONTE CARLO PARTICLE TRANSPORT METHODS

The *history-based* Monte Carlo and the *event-based* Monte Carlo algorithms are inherently parallel because *histories/events* are independent of each other; therefore, they can be processed on separate processors. This means that the development of a parallel Monte Carlo code is straightforward, especially in a distributed computing environment or on a distributed- memory MIMD machine.

Any *history-based* Monte Carlo code can be easily parallelized by considering the following three major steps:

1. All processors are initialized.

2. Particle sources, i.e., histories are distributed evenly among processors and a communication interval for estimating tallies and their uncertainties is set.

3. In case of convergence, all the processors are terminated.

Generally, a software framework can be developed to perform the above three steps. Some effort is needed to decide on the frequency of communication among processors. This issue is more complicated for an eigenvalue problem. Commonly, a fixed-source Monte Carlo simulation can achieve near linear speedup, while the performance for an eigenvalue problem is diminished because of the need for communication and synchronization after every cycle/generation.

12.5.1 Other possible parallel Monte Carlo particle transport algorithms

As discussed, parallel Monte Carlo algorithms based on processing histories/events on different processors are straightforward; however, they require that the entire geometry and material data should be transferred to all processors. This fact may limit the use of such implementations if the available processors do not have sufficient memory. To address this issue, one may develop new algorithms based on the *domain decomposition* techniques. For example, one may partition the spatial domain into subdomains, and process each or groups of subdomain(s) on different processors. Each processor only needs data associated with its assigned subdomain(s). This approach may be more useful for eigenvalue (criticality) calculations, and it may suffer from increased communications. Further, its implementation is significantly more complicated than the *history-based* parallelization.

12.6 DEVELOPMENT OF A PARALLEL ALGORITHM USING MPI

As mentioned earlier, there are different types of parallel architectures; however, the distributed-memory architecture is the most common. To develop a parallel algorithm for a distributed-memory computer, most often, two libraries of functions/routines are used. These are Parallel Virtual Machine (PVM) and Message Passing Interface (MPI). Here, we focus on the MPI library; the PVM library is very similar.

A parallel algorithm conducts five major functions:

1. Initialization

2. Communication

3. Computation

4. Synchronization

5. Termination

To accomplish these functions, MPI's reserved parameters have to be declared by including the *use mpi* instruction, and only a few subroutines have to be invoked for developing a parallel algorithm. Interested readers should consult [42] to learn about the MPI functions or subroutines.

As mentioned, a *history-based* Monte Carlo simulation is embarrassingly parallel, so it is possible to develop a simple framework for parallelizing any Monte Carlo code. Figures 12.3 and 12.4 provide the *mpitest* program; *mpitest* is a framework for parallelizing any Monte Carlo code and includes the five components of a parallel algorithm. Note that communication and synchronization is performed by a *collective* routine that combines the particle counts from all processors. Further, a *master* processor, with zero rank, sets the number of particle histories for each processor, determines the uncertainties of the counts, makes a decision on the termination of a simulation, and prints inputs and outputs.

In performing a parallel Monte Carlo, it is necessary that each processor use a unique seed for its random number generator. A dedicated parallel system should be used to achieve the best performance. To examine the parallel performance, one has to be realistic about the expected performance; a good indicator is the estimate of the parallelization fraction of the code using the Amdahl's law.

12.7 REMARKS

This chapter discussed the concepts of vector and parallel processing. Event-based vector Monte Carlo algorithms were introduced and the difficulties of vectorizing the conventional history-based algorithms were discussed. History-based and event-based parallel Monte Carlo formulations, and their implementation, especially in distributed computing and distributed-memory MIMD environments, were discussed. Effective utilization of the Monte Carlo method for complex physical problems, e.g., nuclear reactors, is only realized if one exploits the available parallel processing capabilities. Finally, a sample framework that can be used for parallelization of any serial Monte Carlo code is provided.

```
program mpitest
*********************************************************************
*Developed by Katherine Royston, PhD Candidate in Nuclear Eng. At VT, 2014
*********************************************************************
*Declaration of mpi-specific parameters
*********************************************************************
use mpi
use transportData
IMPLICIT NONE
integer :: ierr, my_rank, num_proc, seed
integer, parameter :: nMax = 10000000
integer, parameter :: nInc = 1000000
real :: t_start, t_finish
integer :: totTrans, totRefl, totAbs, numTot
integer :: tmpTrans, tmpRefl, tmpAbs
real :: R(3)
*********************************************************************
*Initialization
*********************************************************************
 call mpi_init(ierr)
call mpi_comm_size(MPI_COMM_WORLD,num_proc,ierr)
call mpi_comm_rank(MPI_COMM_WORLD,my_rank,ierr)
*********************************************************************
* Master processor starts the timer and prints the parameters
if(my_rank .EQ. 0) then
   call cpu_time(t_start)
   write(*,*) 'An MPI example of Monte Carlo particle transport.'
   write(*,*) 'A particle is transported through a 1-D shield.'
   write(*,*) ' '
   write(*,*) 'The number of processes: ', num_proc
   write(*,*) ' '
endif
* Make a seed for random number generator on each processor using processor ID
seed = 123456789 + my_rank*10000
call srand(seed)
R(:) = 1.0
numTot = 0
totTrans = 0
totRefl = 0
totAbs = 0
do while (maxval(R) .GT. 0.1 .AND. numTot .LT. nMax)
   numTot = numTot+nInc
*********************************************************************
*Computation – call a Monte Carlo code
*********************************************************************
call transport(nInc/num_proc)
```

Figure 12.3 A framework for parallelization of a Monte Carlo code using MPI

```
*****************************************************************
*Communication & Synchronization
*****************************************************************
   call MPI_Reduce(numTrans, tmpTrans, 1, MPI_INTEGER, MPI_SUM, 0,
MPI_COMM_WORLD, ierr)
   call MPI_Reduce(numRefl, tmpRefl, 1, MPI_INTEGER, MPI_SUM, 0,
MPI_COMM_WORLD, ierr)
   call MPI_Reduce(numAbs, tmpAbs, 1, MPI_INTEGER, MPI_SUM, 0,
MPI_COMM_WORLD, ierr)
*****************************************************************
   totTrans=totTrans+tmpTrans
   totRefl=totRefl+tmpRefl
   totAbs=totAbs+tmpAbs
   R(1)=sqrt((1/real(totTrans))-(1/real(numTot)))
   R(2)=sqrt((1/real(totRefl))-(1/real(numTot)))
   R(3)=sqrt((1/real(totAbs))-(1/real(numTot)))
   if(my_rank .EQ. 0) then
      write(*,*) 'R:', R
      write(*,*) 'Total histories simulated:', numTot
   endif
enddo
if(my_rank .EQ. 0) then
   write(*,'(a)') ' Transmitted Reflected Absorbed'
   write(*,'(a,i13,i13,i13)') 'Number of particles: ', totTrans, totRefl, totAbs
   write(*,'(a,e14.5,e14.5,e14.5)') 'Percent of particles: ', 100.0*
        real(totTrans)/real(numTot), &
        100.0*real(totRefl)/real(numTot), 100.0*real(totAbs)/
        real(numTot)
   write(*,'(a,e14.5,e14.5,e14.5)') 'Relative error: ', R(1), R(2), R(3)
   call cpu_time(t_finish)
   write(*,'(a,f10.5,a)') 'Computation time:', t_finish-t_start, 'sec'
endif
*****************************************************************
*Termination
*****************************************************************
call mpi_finalize(ierr)
end program
```

Figure 12.4 A framework for parallelization of a Monte Carlo code using MPI (continued)

PROBLEMS

1. Table 12.2 gives the speedup achieved by a parallel code for different numbers of processors. Determine the parallel fraction of this code using Equation 12.7.

2. Write a simple parallel code:

   a. Using the following set of instructions given in FORTRAN:

```
program hello
    include 'mpif.h'
    integer rank, size, ierror, tag,
    status(MPI$_STATUS_SIZE)
    call MPI_INIT(ierror)
    call MPI_COMM_SIZE(MPI_COMM_WORLD,
    size, ierror)
    call MPI_COMM_RANK(MPI_COMM_WORLD,
    rank, ierror)
    print*, 'node', rank, ': Hello
    world'
    call MPI_FINALIZE(ierror)
end
```

   b. Modify your program such that every processor says "hello" to the other processors. (Hint: use MPI`send and MPI`receive subroutines).

Table 12.2 Speedup for varying number of processors

# Processors	Speedup
2	1.90
3	2.64
4	3.26
6	4.78
8	4.30
9	5.49
12	6.24
16	6.14
18	8.74
24	7.26
27	8.74

3. Write a parallel program to solve:

$$I = \int_1^2 dx \; ln(x)$$

based on the deterministic Simpson's rule that is discussed as follows. Given $f(x) = ln(x)$, N is the number of intervals considered in the integral range $[a, b]$, and the size of each interval is $h = \frac{b-a}{N}$, then the above integral is obtained by solving:

$$I = \sum_{i=1}^N \frac{h}{3} \left[\left(x_i + \frac{h}{2} \right) + 4f(x_i) + f\left(x_i - \frac{h}{2} \right) \right]$$

Note that you need to start with a guess for N and compare the evaluated solution with the analytical solution. If the relative difference is more than 0.01%, then increase N until this tolerance is satisfied.

4. Write a parallel Monte Carlo program to solve the integral in Problem 12.3 by using:

    a. A uniform *pdf*.

    b. The importance sampling method (see Chapter 5).

   Compare the performance with Problem 10.3.

5. Write a parallel Monte Carlo program to solve:

$$I = \int_{-3}^3 dx \; (1 + tanh(x))$$

    a. Use the stratified sampling method.

    b. Solve the integral using your programs from Problems 12.2 and 12.3.

    c. Compare the results from parts (a) and (b).

6. Using MPI, parallelize your code from Problem 7.1. Run the parallel code on different numbers of processors and discuss your results by evaluating the parallel fraction of your code using Equation 12.7.

7. Write a parallel event-based Monte Carlo algorithm for a 1-D, multi-region slab. Test your code as in Problem 6.4.

Appendix 1

CONTENTS

A.1   INTEGER OPERATIONS ON A BINARY COMPUTER

Some software, such as Matlab and MS-Excel, by default assumes that all variables are floating-point numbers. This can be troublesome in the case of random number generators because the theory behind congruential generators is based on integer arithmetic. A standard 32-bit, single-precision, floating point number uses 23 bits to express the significant digits, with 8 bits for the exponent and 1 bit for sign. A standard 32-bit integer uses 31 bits to store the number and 1 bit for sign. Hence, when dealing with large integer numbers, the floating point representation is "missing" information in the least significant digits. This can cause significant errors, for example, when the modulo function is used. In order to develop a congruential generator that is supposed to work with integer variables, it is necessary that the users declare the variables as integers. If this is not done, then the resulting generator can yield a biased sequence.

The following example shows the representation of integer numbers on digital computers, using binary notation. Suppose we have a computer that can store an 8-bit integer. The largest number, in decimal notation, that we can store is 2^8-1. In binary notation this number can be represented by Figure A.1, where the MSB is the Most Significant Bit.

If we perform a binary addition to the previous number, we obtain Figure A.2. In the result in Figure A.3, we observe a shifting on the left of 1 bit, increasing the bits required to represent the number from 8 to 9.

1	1	1	1	1	1	1	1

MSB

Figure A.1   Binary notation for an 8-bit integer

7	6	5	4	3	2	1	0	Bit #
1	1	1	1	1	1	1	1	
0	0	0	0	0	0	0	1	+

Figure A.2   Binary addition

However, if the computer cannot store this bit in Figure A.3, then integer overflow occurs and the consequences will depend on the particular system and software implementation. If the variable is an unsigned integer, then, generally, this extra bit is simply discarded, and the result in this case for the 8-bit example above yields the number 0. Essentially, every operation is followed by an implicit (mod 2^8). If the variable is a signed integer, other things can happen, such as the number becoming negative. It is very important to know the properties of the particular integer implementation. In the case of creating a congruential random number generator, many of the operations result in integer overflow.

For example, consider again an 8-bit integer, where we desire to use a congruential generatir wit $M = 2^8 - 1$ as given by

$$x_{k+1} = (ax_k + b) mod M \qquad (A.1)$$

If this operation is performed naively, then the operation $(ax_k + 1)$ is performed first. This may result in an integer overflow, but the result has to be stored on 8 bits, so is restricted to the range $[0, 2^8]$. Hence, when the modulo $(2^8 - 1)$ operation is performed, the result is not as

8	7	6	5	4	3	2	1	0	Bit #
1	0	0	0	0	0	0	0	0	

Figure A.3   Result of a binary addition

expected because the number has already been operated on by modulo 2^8

To avoid overflow while performing integer operations, [15] developed the following algorithm. First, new variables q and r are obtained through integer operations

$$q = m \div a \tag{A.2}$$

$$r = (m) \bmod a \tag{A.3}$$

Then, Equation A.1 is replaced with Equation A.4, which is evaluated using integer operations

$$x_{k+1} = a \times [(x_k) \bmod q] - r \times (x_k \div q) \tag{A.4}$$

In Equation A.4, if $x_{k+1} \leq 0$, then

$$x_{k+1} = x_{k+1} + m \tag{A.5}$$

Finally, a new random number is obtained using floating point operations by

$$\eta = \frac{x_{k+1}}{m - 1} \tag{A.6}$$

Appendix 2

CONTENTS

B.1 DERIVATION OF A FORMULATION FOR THE SCATTERING DIRECTION IN A 3-D DOMAIN

Following an elastic scattering interaction, a particle changes its direction ($\hat{\Omega}$) to a new direction ($\hat{\Omega}'$) within $d\Omega'$. As discussed in Chapter 6, Section 6.3.3, by sampling the differential scattering cross-section, the scattering angle, i.e., θ_0 is determined. Then, the direction cosines for the new direction are determined using formulations that relate these direction cosines to the direction cosines of the original direction and the scattering angle. In this section, we will derive these formulations.

Figure B.1 depicts the scattering angle and particle directions before and after scattering in a system of coordinates with unit vectors $\hat{i}$, $\hat{j}$, and $\hat{k}$.

Let's consider that the original direction is given by

$$\hat{\Omega} = u\hat{i} + v\hat{j} + w\hat{k} \tag{B.1}$$

where

$$u = sin\theta cos\phi$$

$$v = sin\theta sin\phi$$

$$w = cos\theta$$

where θ is the polar angle and ϕ is the azimuthal angle.

The direction after scattering is given by

$$\hat{\Omega}' = u'\hat{i} + v'\hat{j} + w'\hat{k} \tag{B.2}$$

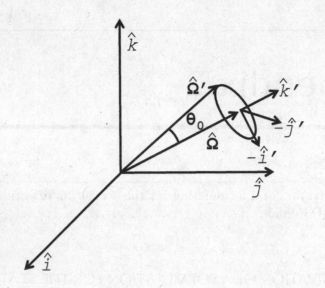

Figure B.1 Schematic of directions before and after scattering

Now, we have to derive formulations for u', v', and w' in terms of u, v, and w, and the scattering angle (θ_0, ϕ_0). Let's define a new system of coordinates $\hat{i}'$, $\hat{j}'$, and $\hat{k}'$, where $\hat{k}'$ is along the direction ($\hat{\Omega}$) before scattering; hence,

$$\hat{k}' = \hat{\Omega}, \tag{B.3}$$

$$\hat{j}' = \frac{\hat{\Omega} \times \hat{k}}{|\hat{\Omega} \times \hat{k}|}, \tag{B.4}$$

and

$$\hat{i}' = \hat{j}' \times \hat{k}' \tag{B.5}$$

Now, we derive the formulations of the unit vectors, $\hat{i}'$, $\hat{j}'$, and $\hat{k}'$, as follows:

$$\hat{j}' = \frac{\hat{\Omega} \times \hat{k}}{|\hat{\Omega} \times \hat{k}|} =$$
$$\left(\frac{v}{s}\right)\hat{i} - \left(\frac{u}{s}\right)\hat{j} \tag{B.6}$$

where $s = \sqrt{1 - w^2}$ (considering $u^2 + v^2 + w^2 = 1$).

$$\hat{i}' =$$
$$\left(\frac{-uw}{s}\right)\hat{i} + \left(\frac{-uw}{s}\right)\hat{j} + s\hat{k} \tag{B.7}$$

Then, based on the new system coordinates, $\hat{\Omega}'$ si given by

$$\hat{\Omega}' = (sin\theta_0 cos\phi_0)\hat{i}' + (sin\theta_0 sin\phi_0)\hat{j}' + (cos\theta_0)\hat{k}' \qquad (B.8)$$

Now, we obtain the direction cosines of $\hat{\Omega}'$ relative to the original system of coordinates as follows:

$$u' = \hat{i} \cdot \hat{\Omega}' = (sin\theta_0 cos\phi_0)\hat{i} \cdot \hat{i}' + (sin\theta_0 sin\phi_0)\hat{i} \cdot \hat{j}' + (cos\theta_0)\hat{i} \cdot \hat{k}' \quad (B.9)$$

$$v' = \hat{j} \cdot \hat{\Omega}' = (sin\theta_0 cos\phi_0)\hat{j} \cdot \hat{i}' + (sin\theta_0 sin\phi_0)\hat{j} \cdot \hat{j}' + (cos\theta_0)\hat{j} \cdot \hat{k}' \quad (B.10)$$

$$w' = \hat{k} \cdot \hat{\Omega}' = (sin\theta_0 cos\phi_0)\hat{k} \cdot \hat{i}' + (sin\theta_0 sin\phi_0)\hat{k} \cdot \hat{j}' + (cos\theta_0)\hat{k} \cdot \hat{k}' \quad (B.11)$$

To complete this derivation, we have to find formulations of the nine scalar products of the unit vectors of the two coordinate systems. These products are determined as follows:

$$\hat{i} \cdot \hat{i}' = \left(\frac{-uw}{s}\right) \qquad \hat{i} \cdot \hat{j}' = \left(\frac{v}{s}\right) \qquad \hat{i} \cdot \hat{k}' = u \qquad (B.12)$$

$$\hat{j} \cdot \hat{i}' = \left(\frac{-vw}{s}\right) \qquad \hat{j} \cdot \hat{j}' = \left(\frac{-u}{s}\right) \qquad \hat{j} \cdot \hat{k}' = v \qquad (B.13)$$

$$\hat{k} \cdot \hat{i}' = s \qquad\qquad \hat{k} \cdot \hat{j}' = 0 \qquad\qquad \hat{k} \cdot \hat{k}' = w \qquad (B.14)$$

Now, if we substitute for the above nine products into equations of the direction cosines, then the direction cosines of the direction following scattering are given below:

$$u' = \hat{i} \cdot \hat{\Omega}' = -\left(\frac{uw}{s}cos\phi_0 - \frac{v}{s}\right)sin\theta_0 + ucos\theta_0, \qquad (B.15)$$

$$v' = \hat{j} \cdot \hat{\Omega}' = -\left(\frac{vw}{s}cos\phi_0 + \frac{u}{s}\right)sin\theta_0 + vcos\theta_0, \qquad (B.16)$$

$$w' = \hat{i} \cdot \hat{\Omega}' = s(sin\theta_0 cos\phi_0) + wcos\theta_0, \qquad (B.17)$$

where $s = \sqrt{1 - w^2}$.

Discussion on the 1-D geometry Note that in a 1-D simulation, commonly we align the z-axis along the 1-D model, then any direction can be recognized by its direction cosine onto the z-axis, i.e., simply Equation B.17. We may rewrite this equation as follows:

$$w' = wu_0 + \sqrt{1 - w^2}\sqrt{1 - \mu_0^2} \times cos\phi_0, \tag{B.18}$$

where $\mu_0 = cos\theta_0$.

Commonly for w and w', we use μ and μ', hence the above equation reduces to:

$$\mu' = \mu\mu_0 + \sqrt{1 - \mu^2}\sqrt{1 - \mu_0^2} \times cos\phi_0 \tag{B.19}$$

Note that the above equation is identical to Equation 6.2 introduced in Chapter 6.

Appendix 3

CONTENTS

C.1   SOLID ANGLE FORMULATION

A solid angle $(d\Omega)$ refers to the angle subtended by area dA_r (normal to radius $\vec{r}$) shown in Figure C.1 and divided by the square of the radius as follows:

$$d\Omega = \frac{dA_r}{r^2} \qquad (C.1)$$

As shown in Figure C.1, dA_r in the spherical system of coordinates is given by

$$dA_r = r^2 sin\theta d\theta d\phi \qquad for\ 0 \le \theta \le \pi,\ and\ 0 \le \phi \le 2\pi \qquad (C.2)$$

Then, the solid angle is given by

$$d\Omega = \frac{r^2 sin\theta d\theta d\phi}{r^2} = sin\theta d\theta d\phi \qquad (C.3)$$

Now, if we consider a change of variable of $\zeta = -cos\theta$, then

$$d\zeta = -sin\theta d\theta \qquad (C.4)$$

Therefore, $d\Omega$ reduces to

$$d\Omega = -d\mu d\phi \qquad (C.5)$$

Since θ varies in the range of $[0, \pi]$, μ varies in a range of $[1, -1]$. If we consider another change of $\mu = -\zeta$, then $d\mu = -d\zeta$, and therefore, μ varies in a range of $[-1, 1]$. Hence $d\Omega$ reduces to

$$d\Omega = d\mu d\phi, \qquad for\ -1 \le \mu \le 1,\ and\ 0 \le \phi \le 2\pi \qquad (C.6)$$

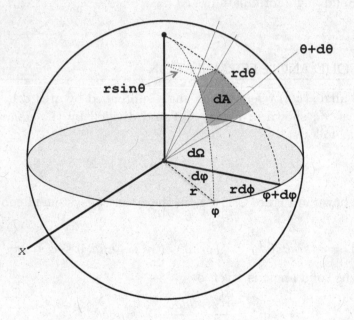

Figure C.1   Schematic of of a solid angle in a spherical system coordinate

Appendix 4

CONTENTS

D.1   ENERGY-DEPENDENT NEUTRON-NUCLEAR INTERACTIONS IN MONTE CARLO SIMULATION

D.2   INTRODUCTION

In this book, for simplicity, we have assumed that a neutron does not change its energy following an interaction with a nuclide in a medium. Further, we have considered only two types of interactions, i.e., scattering and absorption (capture), while, in general, a particle may undergo different types of interactions. In this appendix, for completeness, we briefly address these issues. For further discussions, the reader should consult the MCNP5 manual [96, 67, 94].

The major neutron-nucleus interactions are listed below:

1. Scattering

    (a) elastic (n, n)

    (b) inelastic (n, n')

2. Absorption

    (a) capture (n, γ)

    (b) (n, 2n)

  (c) charged particle (n, p), (n, α)

  (d) fission (n, f)

To account for the variations of cross sections with energy, Monte Carlo codes are designed to use multigroup and/or continuous energy cross-section libraries. The most detailed, i.e., continuous cross-section library in the United States is the ENDF/B-VIII(Evaluated Nuclear Data File, Version VIII [16] library.

In Chapter 6, we discussed how the scattering angle is sampled in elastic scattering. Here, we will develop formulations for sampling the neutron energy and angle following different interactions. Note that the $(n, 2n)$ interaction is not discussed because its cross-section is very small for reactor applications, and, therefore, it is combined with inelastic scattering.

D.3   ELASTIC SCATTERING

To determine energy and direction of motion of a neutron following an elastic scattering, we may utilize different forms of the differential scattering cross-section to sample the scattering angle and neutron energy.

The **scattering angle** (μ_{cm}) in the center-of-mass (CM) system is determined using the tabulated differential scattering cross-section, $\sigma_s(E, \mu_{cm})$. The fundamental formulation of Monte Carlo (FFMC), Equation (2.15) is formed as

$$P(E_k, \mu_{cm}) = \eta$$

$$2\pi \int_{-1}^{\mu_c} d\mu_c \frac{\sigma_s(E_k, \mu_{cm})}{\sigma_s(E_k)} = \eta, \tag{D.1}$$

where E_k refers to a discrete energy in k^{th} energy bin. Equation D.1 is solved for μ_c. In ENDF libraries, when μ_{cm} polynomial has order > 3, the Newton–Raphson method is used. Commonly, however, rather than solving Equation D.1, tables of $P(E_k, \mu_{cm})$ are prepared and used. For a table with J entries, we may consider the following steps:

  1. Generate a random number, η;

  2. Sample the scattering table by using $j = INT(J \cdot \eta)$.

3. Determine deviation from an actual table entry using, $d = \eta \cdot J - j$.

4. Calculate $\mu_c = \mu(k, j) + d(\mu(k, j+1) - \mu(k, j))$.

Neutron energy following the elastics scattering process is deter-mined via kinetics equations. In the lab system, the scattering angle is give by

$$\mu_0 = \frac{1}{2}\left[(A+1)\sqrt{\frac{E'}{E}} - (A-1)\sqrt{\frac{E}{E'}}\right] \tag{D.2}$$

where E and E' refer to the particle energy before and after scattering and A is the ratio of mass of nucleus to the mass of neutron.

The neutron energy after scattering is given by

$$E' = \frac{E}{(A+1)^2}\left[\mu_0 + \sqrt{\mu_0^2 + A^2 - 1}\right]^2 \tag{D.3}$$

D.4   INELASTIC SCATTERING

In inelastic scattering, besides the change in particle direction and kinetic energy, some of the energy is consumed in excitation energy of the nucleus prior to sampling the direction and energy of the scattered neutron. An inelastic scattering may occur if the particle energy is at least greater than the first excited level of the nucleus. The procedure for determining the particle energy and direction following inelastic scattering is given below:

Excitation level is determined by

$$\eta = \sum_i^\infty p(E_k, E_i) \tag{D.4}$$

where m refers to the mth excitation level. $P(E_k, E_i)$ is given by

$$p_k(E_k, E_i) = \frac{\Sigma_{s,n'}(E_k, E_i)}{\Sigma_{s,n'}(E_k)} \tag{D.5}$$

where $\Sigma_{s,n'}(E_k)$ refers to the probability per unit length of inelastic scattering, and $\Sigma_{s,n'}(E_k, E_i)$ refers to the probability per unit length that the residual nucleus will be excited to the ith level. The scattering angle in the CM system (μ_{cm}) is sampled from $P(E_k, \mu_{cm})$ as was performed for elastic scattering.

The scattering angle in the LAB system (μ_0) is determined via

$$\mu_0 = \frac{1}{2}\left[(A+1)\sqrt{\frac{E'}{E}} - (A-1)\sqrt{\frac{E}{E'}} - \frac{QA}{\sqrt{EE'}}\right] \qquad \text{(D.6)}$$

and particle energy after scattering is determined by

$$E' = \frac{1}{(A+1)^2}\left[\mu_0\sqrt{E} + \sqrt{E(\mu_0^2 + A^2 - 1) + A(A+1)Q}\right]^2, \qquad \text{(D.7)}$$

where Q refers to the kinetic energy that is retained by the target nucleus. Note that Equations D.6 and D.7 reduce to Equations D.4 and D.5, respectively, if Q is set equal to zero.

For further information on the formulations of sampling the scattering process, the readers should consult [67] and [96].

D.5 SCATTERING AT THERMAL ENERGIES

At thermal energies, less than a few eV, the neutron-nucleus interaction becomes very complex because the neutron effectively interacts with the whole atom. In this situation, the neutron interactions depend on the neutron energy, target motion, nuclear spin, etc. Chemical binding effects cannot be neglected because the thermal neutron energy is comparable to the thermal motion energy of the atoms (or molecules) in the medium. A neutron can gain or lose energy due to atomic or molecular inelastic scattering (caused by excitation or deexcitation of atoms/molecules). Because of these complexities, the double differential thermal scattering cross sections are written as a single function or "scattering law" named $S(\alpha, \beta)$, where α and β correspond to the momentum and energy transfer between the neutron and nucleus.

As an approximate approach, one may use the "free gas" model that neglects the interference between the target atoms or chemical binding. Such a model treats the target atoms with a Maxwellian distribution given by

$$f(E) = \left[\frac{2}{\sqrt{\pi}}\frac{\sqrt{E}}{(KT)^{\frac{3}{2}}}\right] e^{-\frac{E}{KT}} \qquad \text{(D.8)}$$

where T is the medium temperature in Kelvin and K is the Boltzmann constant.

Appendix 5

CONTENTS

E.1   SHANNON ENTROPY

Here, we will derive the formulation for the Shannon entropy via two approaches.

E.1.1   Derivation of the Shannon entropy - Approach 1

Consider an experiment in which we randomly pick one object out of N objects with a uniform probability of $(\frac{1}{N})$. The amount of information needed or entropy associated with this experiment is dependent on all possible outcomes, i.e.,

$$S\left[\frac{1}{N}, \frac{1}{N}, \cdots\cdots, \frac{1}{N}\right] \equiv f(N) \qquad (E.1)$$

Now, if we change the experiment by grouping the N objects into m groups, with each group containing n_k objects, then we conduct the random selection in two steps. In the first step, we randomly select one of the groups with the probability of

$$p_k = \frac{n_k}{N}, \qquad for \ k = 1, m \qquad (E.2)$$

In the second step, we randomly pick one object from the selected k^{th} group with the probability of $\tilde{p}_k = \frac{1}{n_k}$. Now, we use the composition law for compound experiments, expressed by

$$S(A|B) = S(A) + \sum_{k=1}^{m} p_k S(B|A), \qquad (E.3)$$

where A refers to the first experiment and B refers to the second experiment. So, for the current application, composition formulation is given by

$$f(N) = S(p_1, p_2, \cdots\cdots, p_m) + \sum_{k=1}^{m} p_k f(n_k), \tag{E.4}$$

where

$$f(n_k) = S\left(\frac{1}{n_1}, \frac{1}{n_2}, \cdots\cdots, \frac{1}{n_m}\right) \tag{E.5}$$

In order to decide on the f($\cdot$) function, let;s consider a special case that all the groups have the same number of objects, i.e.,

$$n_1 = n_2 = \cdots = n_m = n. \tag{E.6}$$

This means that $N = m \times n$, and therefore p_k (Equation E.2) reduces to $\frac{1}{m}$, and hence Equation E.4 reduces to

$$f(N) = S\left(\frac{1}{m}, \frac{1}{m}, \cdots\cdots, \frac{1}{m}\right) + \sum_{k=1}^{m} \frac{1}{m} f(n)$$

$$f(N) = f(m) + f(n) \sum_{k=1}^{m} \frac{1}{m} \tag{E.7}$$

$$f(n \times m) = f(m) + f(n)$$

Now, we solve fo $f(m)$

$$f(m) = f(n \times m) - f(n) \tag{E.8}$$

If we consider that $f - functions$ on the right-hand side of the above equation are logarithmic functions, i.e.,

$$f(r) = C \times log(r), \tag{E.9}$$

where C is an arbitrary constant, and r is a dummy variable. Then, we obtain a formulation for $f(m)$ as follows

$$f(m) = C\left[log(n \times m) - log(n)\right] = C\left[log(n) + log(m) - log(n)\right]$$
$$f(m) = C \times log(m)$$

$$\tag{E.10}$$

This means that the logarithmic function is appropriate for this application. Now, using the above log formulation in Equation E.10, we can obtain a formulation for entropy of groups as follows

$$C \times log(N) = S(p_1, p_2, \cdots, p_n) + \sum_{k=1}^{m} p_k \times C \times log(n_k)$$

$$S(p_1, p_2, \cdots, p_n) = C \times log(N) - \sum_{k=1}^{m} p_k \times C \times log(n_k)$$

(E.11)

If substitute for n_k using Equation E.2, the above equation reduces to

$$S(p_1, p_2, \cdots, p_n) = C \times log(N) - \sum_{k=1}^{m} p_k \times C \times log(N) - \sum_{k=1}^{m} p_k \times C \times log(p_k)$$

$$S(p_1, p_2, \cdots, p_n) = -C \sum_{k=1}^{m} p_k log(p_k)$$

(E.12)

Considering that, in the *information theory*, computer bits are commonly used for storing information, it is appropriate to use the *log* function with *base*2. This means that the above equation can be written as

$$H(p) = -K \sum_{k=1}^{m} p_k log_2(p_k)$$

(E.13)

where $H(p)$ is the *Shannon entropy* associated with a probability density function (p), which is made of a discrete set of probabilities (p_k). Finally, it is instructive to derive the Boltzmann formulation for entropy from the Shannon entropy. Let's consider that all the groups are equally probable, i.e., considering that the number of groups is Ω, then $p_k = \frac{1}{\Omega}$, and therefore, Equation E.13 reduces to

$$H = -K \sum_{k=1}^{\Omega} \frac{1}{\Omega} log_2 \left(\frac{1}{\Omega} \right) = K \times log\Omega.$$

(E.14)

The above equation resembles the Boltzmann equation if replace c withe Boltzmann constant (k_B). Therefore, in the Information Theory, it is stated that the *Boltzmann equation* is a special case of the *Shannon entropy*.

E.1.2   Derivation of the Shannon entropy - Approach 2

In the above example, we question the probability of obtaining a set of group outcomes, i.e.,

$$p \equiv p(n_1, n_2, \cdots, n_m) =?. \qquad \text{(E.15)}$$

where

$$N = \sum_{i=1}^{m} n_i \qquad \text{(E.16)}$$

This is a multinomial probability distribution defined by

$$p = \frac{\Gamma}{T} = \frac{\text{\# of combinations of outcomes, } n_i\text{'s}}{\text{total \# of combinations of outcomes, for set of } m \text{ events}} \qquad \text{(E.17)}$$

or

$$p = \frac{\frac{N!}{n_1! n_2! \cdots n_m!}}{m^n} \qquad \text{(E.18)}$$

or

$$p = \left[\frac{N!}{n_1! n_2! \cdots n_m!} \right] \left(\frac{1}{m} \right)^N \qquad \text{(E.19)}$$

Based on the above discussion in the Approach 1, it was concluded that entropy is related to the *log* of the number of outcomes, i.e.,

$$H = log_2 \Gamma = log_2 \left[\frac{N!}{n_1! n_2! \cdots n_m!} \right] \qquad \text{(E.20)}$$

Here for ease of derivation, we set $log_2 \equiv log$, and expand the right-hand side of the above equation as follows

$$H = log(N!) - log(n_1!) - log(n_2!) \cdots - log(n_m!)$$

$$H = \sum_{i=1}^{N} log(i) - \sum_{i=1}^{n_1} log(i) - \sum_{i=1}^{n_2} log(i) - \cdots - \sum_{i=1}^{n_m} log(i) \qquad \text{(E.21)}$$

If we substitute each sum with an integral using the Stirling's approximation, i.e.,

$$\sum_{i=1}^{k} log(i) = \int_{1}^{k} dx log x = K \times log k - K + 1, \qquad \text{(E.22)}$$

then Equation E.21 reduces to

$$H = (NlogN - N + 1) - (n_1 logn_1 - n_1 + 1) - \cdots$$
$$-(n_m logn_m - n_m + 1)$$

$$H = NlogN - \sum_{x=1}^{m} n_x logn_x + (1 - m)$$

(E.23)

Now, considering that $n_x = Np_x$, the above equation reduces to

$$H = NlogN - \sum_{x=1}^{m} Np_x logNp_x + (1 - m)$$

$$H = NlogN - \sum_{x=1}^{m} n_x(logN + logp_x) + (1 - m)$$

$$H = NlogN - NlogN \sum_{x=1}^{m} p_x - N \sum_{x=1}^{m} p_x logp_x + (1 - m)$$

$$H = -N \sum_{x=1}^{m} p_x log_2 p_x + (1 - m)$$

(E.24)

Note that in the above equation, we have substituted back log_2 for log.

Equation E.24 resembles Equation E.13, but with different constants. In principle, the two equations are equivalent as the constant K in Equation E.13 is an arbitrary parameter. Finally, the *Shannon entropy* is used for examining the relative behavior of fission neutron density from one generation to the next. The parallels to the aforementioned experiment include:

1. N is the total number of fission neutrons.

2. m is the number of subregions with fissile materials in a reactor core.

3. $p_i = p_k$, p_i corresponds to fission density distribution in one generation.

4. Since we are interested in the behavior of entropy from one generation to the next, we have dropped/replaced the constants, and use the following formulation

$$H(p) = -\sum_{i=1}^{m} p_i log_2 p_i.$$

(E.25)

Bibliography

[1] John Allison, Katsuya Amako, J E A Apostolakis, HAAH Araujo, P Arce Dubois, MAAM Asai, GABG Barrand, RACR Capra, SACS Chauvie, and RACR Chytracek. Geant4 developments and applications. *IEEE Transactions on nuclear science*, 53(1):270–278, 2006.

[2] Gene M Amdahl. Validity of the single processor approach to achieving large scale computing capabilities. In *Proceedings of the April 18-20, 1967, spring joint computer conference*, pages 483–485, 1967.

[3] George B Arfken and Hans J Weber. Mathematical methods for physicists, 1999.

[4] George I Bell and Samuel Glasstone. *Nuclear reactor theory*. US Atomic Energy Commission, Washington, DC (United States), 1970.

[5] R N Blomquist. The OECD/NEA source convergence benchmark program. Technical report, 2002.

[6] Roger N Blomquist, Malcolm Armishaw, David Hanlon, Nigel Smith, Yoshitaka Naito, Jinan Yang, Yoshinori Mioshi, Toshihiro Yamamoto, Olivier Jacquet, and Joachim Miss. Source convergence in criticality safety analyses. *NEA Report ISBN 92-64-02304*, 6, 2006.

[7] Roger N Blomquist and Ely M Gelbard. Fission source algorithms and Monte Carlo variances. In *Transactions of the American Nuclear Society*, 2000.

[8] T E Booth. Automatic importance estimation in forward Monte Carlo calculations. *Transactions of the American Nuclear Society*, 41, 1982.

[9] T E Booth. A caution on reliability using optimal variance reduction parameters. *Transactions of the American Nuclear Society*, 66:278–280, 1992.

[10] Thomas E Booth. Weight window/importance generator for Monte Carlo streaming problems. Technical report, 1983.

[11] Thomas E Booth. Monte Carlo variance comparison for expected-value versus sampled splitting. *Nuclear Science and Engineering*, 89(4):305–309, 1985.

[12] Thomas Edward Booth. Sample problem for variance reduction in MCNP. Technical report, 1985.

[13] J P Both, J C Nimal, and T Vergnaud. Automated importance generation and biasing techniques for Monte Carlo shielding techniques by the TRIPOLI-3 code. *Progress in Nuclear Energy*, 24(1-3):273–281, 1990.

[14] G E P Box and Mervin E Muller. A Note on the Generation of Random Normal Deviates, The Annals of Mathematical Statistics, 1958.

[15] Paul Bratley, Bennet L Fox, and Linus E Schrage. *A guide to simulation*. Springer Science & Business Media, 2011.

[16] David A Brown, M B Chadwick, R Capote, A C Kahler, A Trkov, M W Herman, A A Sonzogni, Y Danon, A D Carlson, and M Dunn. ENDF/B-VIII. 0: The 8th major release of the nuclear reaction data library with CIELO-project cross sections, new standards and thermal scattering data. *Nuclear Data Sheets*, 148:1–142, 2018.

[17] Forrest B Brown. A review of Monte Carlo criticality calculations-convergence, bias, statistics. Technical report, 2008.

[18] Forrest B Brown and William R Martin. Monte Carlo methods for radiation transport analysis on vector computers. 1984.

[19] Richard Stevens Burington and Donald Curtis May Jr. *Handbook of probability and statistics, with tables*, volume 77. LWW, 1954.

[20] P. J. Burns. Unpublished notes on random number generators, 2004.

[21] W Cai. Unpublished lecture notes for ME346A Introduction to Statistical Mechanics, 2011.

[22] Edmond D Cashwell and Cornelius Joseph Everett. A practical manual on the Monte Carlo method for random walk problems, 1959.

[23] Steve Chucas, Ian Curl, T Shuttleworth, and Gillian Morrell. PREPARING THE MONTE CARLO CODE MCBEND FOR THE 21ST CENTURY. 1994.

[24] Roger R Coveyou, V R Cain, and K J Yost. Adjoint and importance in Monte Carlo application. *Nuclear Science and Engineering*, 27(2):219–234, 1967.

[25] S N Cramer and J S Tang. Variance reduction methods applied to deep-penetration Monte Carlo problems. Technical report, 1986.

[26] Jack J Dongarra, Cleve Barry Moler, James R Bunch, and Gilbert W Stewart. *LINPACK users' guide*. SIAM, 1979.

[27] Jan Dufek. Accelerated monte carlo eigenvalue calculations. In *XIII Meeting on Reactor Physics Calculations in the Nordic Countries Västers, Sweden*, volume 29, page 30, 2007.

[28] Jan Dufek. Development of new monte carlo methods in reactor physics. *KTH Royal Institute of Technology*, 2009.

[29] William L Dunn and J Kenneth Shultis. *Exploring monte carlo methods*. Elsevier, 2011.

[30] S R Dwivedi. A new importance biasing scheme for deep-penetration Monte Carlo. *Annals of Nuclear Energy*, 9(7):359–368, 1982.

[31] M B Emmett. MORSE Monte Carlo radiation transport code system. Technical report, 1975.

[32] C J Everett and E D Cashwell. A third Monte Carlo sampler. *Los Alamos Report LA-9721-MS*, 1983.

[33] C J Everett, E D Cashwell, and G D Turner. Method of sampling certain probability densities without inversion of their distribution functions. Technical report, 1973.

[34] Michael J Flynn. Some computer organizations and their effectiveness. *IEEE transactions on computers*, 100(9):948–960, 1972.

[35] A. H. Foderaro. A Monte Carlo primer. (Unpublished notes), 1986.

[36] N A Frigerio and N A Clark. Random number set for Monte Carlo computations. Technical report, 1975.

[37] Al Geist, Adam Beguelin, Jack Dongarra, Weicheng Jiang, Robert Manchek, and Vaidyalingam S Sunderam. *PVM: Parallel virtual machine: a users' guide and tutorial for networked parallel computing*. MIT press, 1994.

[38] Eo Mo Gelbard and R E Prael. Monte Carlo Work at Argonne National Laboratory. Technical report, 1974.

[39] James E Gentle. Monte carlo methods. *Random number generation and Monte Carlo methods*, pages 229–281, 2003.

[40] Paul Glasserman. *Monte Carlo methods in financial engineering*, volume 53. Springer Science & Business Media, 2013.

[41] Harold Greenspan, C N Kelber, and David Okrent. Computing methods in reactor physics. 1972.

[42] William Gropp, Ewing Lusk, and Anthony Skjellum. Portable Parallel Programming with the Message-Passing Interface, 1994.

[43] A Haghighat, H Hiruta, B Petrovic, and J C Wagner. Performance of the Automated Adjoint Accelerated MCNP (A3MCNP) for Simulation of a BWR Core Shroud Problem. In *Proceedings of the International Conference on Mathematics and Computation, Reactor Physics, and Environmental Analysis in Nuclear Applications*, page 1381, 1999.

[44] A Haghighat and John C Wagner. Application of A 3 MCNP™ to Radiation Shielding Problems. In *Advanced Monte Carlo for Radiation Physics, Particle Transport Simulation and Applications*, pages 619–624. Springer, 2001.

[45] Alireza Haghighat, Katherine Royston, and William Walters. MRT methodologies for real-time simulation of nonproliferation and safeguards problems. *Annals of Nuclear Energy*, 87:61–67, 2016.

[46] Alireza Haghighat and John C Wagner. Monte Carlo variance reduction with deterministic importance functions. *Progress in Nuclear Energy*, 42(1):25–53, 2003.

[47] Max Halperin. Almost linearly-optimum combination of unbiased estimates. *Journal of the American Statistical Association*, 56(293):36–43, 1961.

[48] Donghao He and William J Walters. A local fission matrix correction method for heterogeneous whole core transport with RAPID. *Annals of Nuclear Energy*, 134:263–272, 2019.

[49] J S Hendricks. A code-generated Monte Carlo importance function. *Transactions of the American Nuclear Society*, 41, 1982.

[50] Hideo Hirayama, Yoshihito Namito, Walter R Nelson, Alex F Bielajew, and Scott J Wilderman. The EGS5 code system. Technical report, 2005.

[51] J Eduard Hoogenboom, William R Martin, and Bojan Petrovic. The Monte Carlo performance benchmark test-aims, specifications and first results. In *International Conference on Mathematics and Computational Methods Applied to*, volume 2, page 15, 2011.

[52] Nicholas Horelik, Bryan Herman, Benoit Forget, and Kord Smith. Benchmark for evaluation and validation of reactor simulations (BEAVRS), v1. 0.1. In *Proc. Int. Conf. Mathematics and Computational Methods Applied to Nuc. Sci. & Eng*, pages 5–9, 2013.

[53] Thomas E Hull and Alan R Dobell. Random number generators. *SIAM review*, 4(3):230–254, 1962.

[54] Gwilym M Jenkins and D G Watts. Spectral Analysis and its Applications Holden-Day, San Francisco. *Spectral analysis and its application. Holden-Day, San Francisco*, 1968.

[55] David Kahaner, Cleve Moler, and Stephen Nash. Numerical methods and software. *Englewood Cliffs: Prentice Hall, 1989*, 1989.

[56] Herman Kahn. Applications of Monte Carlo. Technical report, 1954.

[57] Herman Kahn and Theodore E Harris. Estimation of particle transmission by random sampling. *National Bureau of Standards applied mathematics series*, 12:27–30, 1951.

[58] M H Kalos and P A Whitlock. Monte Carlo Methods: Basics, vol. 1, 1986.

[59] Donald Ervin Knuth. *The art of computer programming*, volume 3. Pearson Education, 1997.

[60] A L'Abbate, T Courau, and E Dumonteil. Monte Carlo Criticality Calculations: Source Convergence and Dominance Ratio in an Infinite Lattice Using MCNP and TRIPOLI4. In *PHYTRA1 Conference, Marrakech, Morocco*, 2007.

[61] A Laureau, M Aufiero, P R Rubiolo, E Merle-Lucotte, and D Heuer. Transient Fission Matrix: Kinetic calculation and kinetic parameters βeff and Λeff calculation. *Annals of Nuclear Energy*, 85:1035–1044, 2015.

[62] Pierre L'Ecuyer. Random numbers for simulation. *Communications of the ACM*, 33(10):85–97, 1990.

[63] Pierre L'Ecuyer. Good parameters and implementations for combined multiple recursive random number generators. *Operations Research*, 47(1):159–164, 1999.

[64] Derrick H Lehmer. Mathematical methods in large-scale computing units. *Annu. Comput. Lab. Harvard Univ.*, 26:141–146, 1951.

[65] Elmer Eugene Lewis and Warren F Miller. Computational methods of neutron transport. 1984.

[66] Jun S Liu. *Monte Carlo strategies in scientific computing.* Springer Science & Business Media, 2008.

[67] I Lux and L Koblinger. Monte Carlo Particle Transport Methods: Neutron and Photon Calculations. 1991.

[68] George Marsaglia. DIEHARD: a battery of tests of randomness. *http://stat. fsu. edu/geo*, 1996.

[69] George Marsaglia. Xorshift rngs. *Journal of Statistical Software*, 8(14):1–6, 2003.

[70] George Marsaglia, Arif Zaman, and Wai Wan Tsang. Toward a universal random number generator. *Stat. Prob. Lett.*, 9(1):35–39, 1990.

[71] V Mascolino, A Haghighat, and N J Roskoff. Evaluation of RAPID for a UNF cask benchmark problem. *EPJ Web of Conferences*, 153:5025, 2017.

[72] Valerio Mascolino and Alireza Haghighat. Validation of the Transient Fission Matrix Code tRAPID against the Flattop-Pu Benchmark. In *Proceedings of the International Conference on Mathematics and Computational Methods applied to Nuclear Science and Engineering (M&C 2019)*, pages 1338–1347, Portland, OR, 2019.

[73] Valerio Mascolino, Anze Pungercic, Alireza Haghighat, and Luka Snoj. Experimental and Computational Benchmarking of RAPID using the JSI TRIGA MARK-II Reactor. In *Proceedings of the International Conference on Mathematics and Computational Methods applied to Nuclear Science and Engineering (M&C 2019)*, pages 1328–1337, Portland, OR, 2019.

[74] Valerio Mascolino, Nathan J Roskoff, and Alireza Haghighat. Benchmarking of the Rapid Code System Using the GBC-32 Cask With Variable Burnups. In *PHYSOR 2018: Reactor Physics Paving The Way Towards More Efficient Systems*, pages 697–708, Cancun, Mexico, 2018.

[75] N Metropolis. The beginning of the Monte Carlo Method. *Los Alamos Science*, 15:125–130, 1987.

[76] Gordon E Moore. Cramming more components onto integrated circuits. *Proceedings of the IEEE*, 86(1):82–85, 1998.

[77] Richard L Morin. *Monte Carlo simulation in the radiological sciences*. CRC Press, 2019.

[78] Scott W Mosher, Aaron M Bevill, Seth R Johnson, Ahmad M Ibrahim, Charles R Daily, Thomas M Evans, John C Wagner, Jeffrey O Johnson, and Robert E Grove. ADVANTG - an automated variance reduction parameter generator. *ORNL/TM-2013/416, Oak Ridge National Laboratory*, 2013.

[79] Stephen K Park and Keith W Miller. Random number generators: good ones are hard to find. *Communications of the ACM*, 31(10):1192–1201, 1988.

[80] Denise B Pelowitz. MCNP6 user's manual version 1.0. *Los Alamos National Security, USA*, 2013.

[81] Douglas E Peplow, Stephen M Bowman, James E Horwedel, and John C Wagner. Monaco/MAVRIC: Computational Resources for Radiation Protection and Shielding in SCALE. *TRANSACTIONS-AMERICAN NUCLEAR SOCIETY*, 95:669, 2006.

[82] L M Petrie and N F Cross. KENO IV: An improved Monte Carlo criticality program. Technical report, 1975.

[83] Lennart Rade and Bertil Westergren. *BETA mathematics handbook*, 1990.

[84] Farzad Rahnema and Dingkang Zhang. Continuous energy coarse mesh transport (COMET) method. *Annals of Nuclear Energy*, 115:601–610, 2018.

[85] Adam Rau and William J Walters. Validation of coupled fission matrix–TRACE methods for thermal-hydraulic and control feedback on the Penn State Breazeale Reactor. *Progress in Nuclear Energy*, 123:103273, 2020.

[86] N Roskoff, A Haghighat, and V Mascolino. Analysis of RAPID Accuracy for a Spent Fuel Pool with Variable Burnups and Cooling Times. In *Proceedings of Advances in Nuclear Nonproliferation Technology and Policy Conference, Santa Fe, NM*, 2016.

[87] N J Roskoff and A Haghighat. Development of a novel fuel burnup methodology using the rapid particle transport code system. In *PHYSOR 2018: reactor physics paving the way towards more efficient systems*, Mexico, 2018. Sociedad Nuclear Mexicana.

[88] Nathan J Roskoff, Alireza Haghighat, and Valerio Mascolino. Experimental and Computational Validation of RAPID. In *Proc. 16th Int. Symp. Reactor Dosimetry*, pages 7–12, 2017.

[89] Francesc Salvat, Jose M Fernandez-Varea, and Josep Sempau. PENELOPE-2006: A code system for Monte Carlo simulation of electron and photon transport. In *Workshop proceedings*, volume 4, page 7. Nuclear Energy Agency, Organization for Economic Co-operation and ..., 2006.

[90] Claude E Shannon. A mathematical theory of communication. *Bell system technical journal*, 27(3):379–423, 1948.

[91] Samuel Sanford Shapiro and Martin B Wilk. An analysis of variance test for normality (complete samples). *Biometrika*, 52(3/4):591–611, 1965.

[92] Hyung Jin Shim and Chang Hyo Kim. Stopping criteria of inactive cycle Monte Carlo calculations. *Nuclear science and engineering*, 157(2):132–141, 2007.

[93] Jonathon Shlens. A light discussion and derivation of entropy. *arXiv preprint arXiv:1404.1998*, 2014.

[94] Jerome Spanier and Ely M Gelbard. *Monte Carlo principles and neutron transport problems*. Courier Corporation, 2008.

[95] Student. The probable error of a mean. *Biometrika*, pages 1–25, 1908.

[96] X-5 Monte Carlo Team. MCNP-A General Monte Carlo N-Particle Transport Code, Version 5, 2003.

[97] Tyler J Topham, Adam Rau, and William J Walters. An iterative fission matrix scheme for calculating steady-state power and critical control rod position in a TRIGA reactor. *Annals of Nuclear Energy*, 135:106984, 2020.

[98] Scott A Turner and Edward W Larsen. Automatic variance reduction for three-dimensional Monte Carlo simulations by the local importance function transform—II: numerical results. *Nuclear science and engineering*, 127(1):36–53, 1997.

[99] Taro Ueki. Information theory and undersampling diagnostics for Monte Carlo simulation of nuclear criticality. *Nuclear science and engineering*, 151(3):283–292, 2005.

[100] Taro Ueki and Forrest B. Brown. Stationarity diagnostics using Shannon entropy in monte carlo criticality calculation I: F test. In *American Nuclear Society 2002 Winter Meeting*, pages 17–21, 2002.

[101] Taro Ueki and Forrest B Brown. Informatics approach to stationarity diagnostics of the Monte Carlo fission source distribution. *Transactions of the American Nuclear Society*, pages 458–461, 2003.

[102] Todd J Urbatsch, R Arthur Forster, Richard E Prael, and Richard J Beckman. Estimation and Interpretation of keff Confidence Intervals in MCNP. *Nuclear technology*, 111(2):169–182, 1995.

[103] Eric Veach and Leonidas J Guibas. Optimally combining sampling techniques for Monte Carlo rendering. In *Proceedings of the 22nd annual conference on Computer graphics and interactive techniques*, pages 419–428, 1995.

[104] John Von Neumann. 13. various techniques used in connection with random digits. *Appl. Math Ser*, 12(36-38):5, 1951.

[105] J C Wagner. *Computational Benchmark for Estimation of Reactivity Margin from Fission Products and Minor Actinides in PWR Burnup Credit Prepared by*. 2000.

[106] John C Wagner. Acceleration of Monte Carlo shielding calculations with an automated variance reduction technique and parallel processing, 1997.

[107] John C Wagner. An automated deterministic variance reduction generator for Monte Carlo shielding applications. In *Proceedings of the American Nuclear Society 12th Biennial RPSD Topical Meeting*, pages 14–18. Citeseer, 2002.

[108] John C Wagner and Alireza Haghighat. Automated variance reduction of Monte Carlo shielding calculations using the discrete ordinates adjoint function. *Nuclear Science and Engineering*, 128(2):186–208, 1998.

[109] John C Wagner, Douglas E Peplow, and Scott W Mosher. FW-CADIS method for global and regional variance reduction of

Monte Carlo radiation transport calculations. *Nuclear Science and Engineering*, 176(1):37–57, 2014.

[110] John C. Wagner, Douglas E Peplow, Scott W Mosher, and Thomas M Evans. Review of hybrid (deterministic/Monte Carlo) radiation transport methods, codes, and applications at Oak Ridge National Laboratory. *Progress in nuclear science and technology*, 2:808–814, 2011.

[111] W J Walters. Application of the RAPID Fission Matrix Methodology to 3-D Whole-Core Reactor Transport. In *Proc. Int. Conf. Mathematics and Computational Methods Applied to Nuclear Science and Engineering (M&C 2017)*, pages 16–20, 2017.

[112] William Walters, Nathan J. Roskoff, and Alireza Haghighat. A Fission Matrix Approach to Calculate Pin-wise 3D Fission Density Distribution. In *Proc. M&C 2015*, Nashville, Tennesse, 2015.

[113] William J Walters, Nathan J Roskoff, and Alireza Haghighat. The RAPID Fission Matrix Approach to Reactor Core Criticality Calculations. *Nuclear Science and Engineering*, pages 1–19, Aug 2018.

[114] Mengkuo Wang. *CAD Based Monte Carlo Method: Algorithms for Geometric Evaluation in Support of Monte Carlo Radiation Transport Calculation*. University of Wisconsin–Madison, 2006.

[115] Alvin M Weinberg and Eugene Paul Wigner. The physical theory of neutron chain reactors, 1958.

[116] Michael Wenner. *Development and analysis of new Monte Carlo stationary source diagnostics and source acceleration for Monte Carlo eigenvalue problems with a focus on high dominance ratio problems. PhD diss.* PhD thesis, University of Florida, Gainesville, 2010.

[117] Michael Wenner and Alireza Haghighat. A generalized KPSS test for stationarity detection in Monte Carlo eigenvalue problems, 2008.

[118] Michael Wenner and Alireza Haghighat. A Fission Matrix Based Methodology for Achieving an Unbiased Solution for Eigenvalue Monte Carlo Simulations. *Progress in Nuclear Science and Technology*, 2:886–892, 2010.

[119] Michael T Wenner and Alireza Haghighat. Study of Methods of Stationarity Detection for Monte Carlo Criticality Analysis with KENO Va. *Transactions*, 97(1):647–650, 2007.

[120] G A Wright, E Shuttleworth, M J Grimstone, and A J Bird. The status of the general radiation transport code MCBEND. *Nuclear Instruments and Methods in Physics Research Section B: Beam Interactions with Materials and Atoms*, 213:162–166, 2004.

[121] Charles N Zeeb and Patrick J Burns. Random number generator recommendation. *Report prepared for Sandia National Laboratories, Albuquerque, NM. Available as a WWW document., URL= http://www. colostate. edu/ pburns/monte/documents. html*, 1997.

Index

Printed in the United States
by Baker & Taylor Publisher Services

Printed in the United States
by Baker & Taylor Publisher Services